THERAPEUTIC STORYTELLING
100 Stories for Turbulent Times

Dr PJ Kochuthresia

Dr PJ Kochuthresia, an accomplished educator formerly at Little Flower College, Guruvayur, seamlessly integrates her expertise as a Consultant Psychologist, Performance Coach, and Psychotherapist, specializing in EMDR and behaviour modification. An acclaimed author of multiple books, she uniquely combines psychology and storytelling, offering transformative insights that enrich the craft. Immerse yourself in her captivating narratives, where each tale reflects the profound connection between the human experience and the art of storytelling.

THERAPEUTIC STORYTELLING

100 Stories for Turbulent Times

Dr PJ Kochuthresia

H & C BOOKS

English Language
THERAPEUTIC STORYTELLING
Self-help
Dr PJ Kochuthresia

First Published: January 2024
Cover: Jijoy Johny T.
Pagesetting: Art N Colour
Printed in India
at Ebenezer Printpack, Thrissur

Publishers
H&C PUBLISHING HOUSE
Thrissur - 680 001
Website: www.handcbooks.com
E-mail: handcbooks@gmail.com

Distributors
H&C STORES
Thiruvananthapuram Bhadra Center, Vanchiyoor, Kollam Jerome Nagar, Vadayattukotta, Pathanamthitta, Thiruvalla, Kottayam, Pala,Thodupuzha, Ernakulam SC Rd, Near TDM Hall, Kadavanthra, Kakkanad, Palarivatom, Edappally, Edappally WS, Thrippunithura, Thrissur High Rd, High Rd New, South Bazar, Round North, Ikkanda Warrior Rd, Kunnamkulam New Bus stand Rd, Wadakkanchery Rd, Palakkad, Kozhikode, Kannur, Kanhangad, Bengaluru

ISBN 978-81-19331-58-1
₹ 300.00
24010-500-01-0124

ACKNOWLEDGMENT

I extend my heartfelt gratitude to those whose unwavering support, love, and encouragement have been instrumental in bringing this book to light:

My Beloved Husband, Mathew: Your constant support, love, and inspiration have propelled me to greater heights. Your unwavering belief in me fuels my passion for academics and writing.

My Son, Ashish, His Wife Ann, and Little Ezra: Your presence and steadfast support have been a continual source of strength. Thank you for always being there for me.

My Brother, Dr P. J. James S.J.: Your guidance, valuable suggestions, and meticulous review have significantly enriched the content of this book.

My Brother, Dr Francis Valloor: Your insightful opinions and reassurance have motivated me to strive for excellence.

Dear Friend, Colleague, and Mentor, Dr Usha Balakrishnan: Your support from the inception to completion of this book has been invaluable. Your wisdom and guidance have shaped the course of this project.

Long-time Friend, Dr Anuradha Sunil: Thank you for supporting all my creative eccentricities and being a constant source of encouragement.

Special Thanks to Gauri, Our Niece: Your unwavering encouragement throughout this endeavour is deeply appreciated.

Dr PJ Kochuthresia

Foreword

I have had the privilege of knowing Dr PJ Kochuthresia for many years, both casually and professionally. Our collaborative efforts have not only enriched our own lives but have also positively impacted so many students we have collectively nurtured, mentored, and guided towards healing and fulfilment. When Dr Kochuthresia approached me to pen the foreword for this collection of short and captivating stories, I felt deeply honoured and touched.

Storytelling is an inherent trait that all of us possess, whether we acknowledge it or not. We all share stories - realistic and truthful narratives from our lives, often sprinkled with personal embellishments of adventure, magic, and other fascinating elements. The greater our emotional investment in these stories, the more compelling and instructive they become, passing on valuable life lessons to those who listen or read. These stories serve as examples or cautionary tales, shaping the listener or reader's journey through life.

Dr Kochuthresia's stories transport readers to immersive worlds. Her narratives are unusually concise yet highly educational, some even taking the form of parables. Through relatable characters and engaging situations, she crafts stories that captivate the reader's attention and imagination. Each story, be it for enjoyment, learning, reflection, relaxation, comfort, or guidance, prompts readers to delve into its pages, leaving them pondering the writing style and the lessons gleaned.

Within the pages of this book, you will discover a hundred assorted stories, each crafted with commendable imagination. I am confident that every reader will find these stories as intriguing and instructive as I have. I extend my best wishes for the successful launch and wide-reaching impact of this book!

Dr E. Anuradha Sunil
Principal
Royal Dental College
Chalissery

Preface

Unveiling the Healing Power of Stories

Welcome to a transformative journey, where the art of storytelling takes centre stage as a powerful agent of healing. In the following pages, we embark on an insightful exploration of the profound impact stories wield on the human spirit - of their ability to transcend time, age, and circumstance.

As a species, we are born storytellers. From the lullabies that cradle us in infancy to the sagas that accompany us through the trials of adulthood, stories are the threads that bind the fabric of our lives. In this collection, we delve into the universal language of storytelling, discovering its therapeutic essence.

Stories are more than mere words on a page; they are bridges that connect us. In this book, you will encounter narratives from the far reaches of the world, each resonating with themes of collaboration and connection. Whether you find solace in the simplicity of childhood fables or draw strength from tales of resilience in the face of life's complexities, these stories are crafted to be your companions.

The societal investment in storytelling becomes a focal point of our inquiry, as we navigate the layers of our shared human experience. From our collective willingness to accept things beyond the ordinary to the profound impact stories exert on our emotions and perceptions, we systematically unveil the enchantment intrinsic to the craft of narrative creation.

Consider this book as a repository - a gift of stories passed down through generations. Just as we nurture our bodies, our minds crave the sustenance of narratives. Stories constitute a timeless inheritance, an ancient tradition that transcends cultural confines, fostering connections within families and communities.

At the heart of this collection lies the therapeutic power of storytelling. Stories, in their various forms, serve as catalysts for emotional, cognitive, and linguistic development. They inspire reflection, ignite curiosity, and provide a pathway to understanding life's complexities. This book is an exploration of how stories, carefully chosen and shared, can be agents of healing offering solace and fostering resilience in the face of life's challenges.

This journey invites you to embrace the richness of storytelling, connecting with narratives that resonate across cultures and generations. Let the healing power of these stories unfold as you turn the pages, weaving new threads into the tapestry of your own story.

The stories in this book will soon be published in two volumes: *Healing Stories* and *Empowering Stories*. Immerse yourself in narratives that traverse diverse societies and epochs, inviting you to a captivating realm of vibrant storytelling.

CONTENTS

PART - 1

Stories for Children and for the Child in Every Adult

"We are, as a species, addicted to story. Even when the body goes to sleep, the mind stays up all night, telling itself stories." Jonathan Gottschall

Who is this book for? In fact, it is a puzzling question to answer because as a storyteller I have seen adults enjoying stories written for children and children enjoying stories written for older people. So, this book is for children and for the child in every adult. Storytelling is not just about children or about imagination. Only the nature of the story changes as we grow up. In childhood the stories may be plain. But as we grow older, it may be about breakups, sorrows, defeats, or victory. It is in our nature as social beings to tell and listen to tales whatever be our age. Storytelling is woven deep into the human psyche.

In this collection, we have a treasure trove of stories from different corners of the world. These stories revolve around the themes of collaboration and connection. They have a timeless quality that resonates with people of all ages. These narratives stir our emotions and create a bond between us and others. Stories serve as a medium to share our passions, sorrows, challenges, and moments of happiness. They give our lives meaning and purpose. These stories both enlighten and entertain us, providing a source of healing and completeness.

Stories serve as a universal bridge that helps people connect, breaking down our defences, barriers, and differences. They offer a deeper understanding of ourselves and our commonalities with others. When we listen to a story, it becomes our own in a unique way. When we retell it, we add our own personal touches, circumstances, and may be a bit of exaggeration, while preserving the core plot and characters. It takes on new shades and becomes a part of our listeners' experiences. Stories are shared and passed down, with each person interpreting them through their own perspectives and beliefs.

Stories are universally appealing because they engage our emotions, allowing us to think from the heart rather than the head. Imagination sparks our feelings, and stories serve as mirrors of both the facts and emotions in our lives. We find reflections of our own experiences in them. Listeners or readers immerse themselves in the scenes with the characters, connecting their own life experiences to these narratives. What is more, the finest stories guide us in comprehending life's complexities. Crucially, they can inspire hope, impart moral lessons, emphasize social responsibility, and cultivate a sense of tenderness in our hearts.

Why do we invest so much time telling stories and listening to them that both the teller and listener clearly know never happened or are never likely to happen? This phenomenon is what S.T. Coleridge calls "willing suspension of disbelief," the great "ability of humans to think in sustained fashion beyond the here and now" (Boyd), believing the unbelievable. Storytelling, whether factual or fictional, is a human characteristic. When we are emotionally engrossed in a story, our natural scepticism diminishes, and as a result we drop our mistrust. Stories ignite our imagination and listeners relish the stimulation of their creative faculties. The art of storytelling lies in the mental images that our minds invent while listening. Stories, metaphors, myths, legends, and fairy tales serve various functions, such as passing down family and cultural wisdom, expressing emotions, sharing personal experiences, and conveying values and knowledge.

There are numerous reasons why people find stories so appealing. Stories prompt listeners to reflect on deep issues of life, by blending reason with an element of mystery. They encourage us to perceive the world in a holistic manner, ultimately contributing to healing and a sense of wholeness. Moreover, stories make the learning process exciting. When talented storytellers craft context-specific narratives that resonate with the audience's culture, they are most effective in an informal setting, speaking the language and using the medium familiar to the audience. This approach forges a personal bond between the storyteller and the audience.

The Gift of Stories

Storytelling is an ancient and valuable human treasure. Our vast collection of stories is akin to a rich inheritance. Just as we require nourishment for our bodies, we also crave stories to feed our minds. Stories nourish us and provide a deeper understanding of the world. Whether intentionally or unwittingly, we constantly share tales with our family members, friends, and colleagues all the time. Stories possess an unparalleled power to foster trust and bring about change within individuals. It is safe to say that there is hardly anyone who does not relish the pleasure of listening to and enjoying stories from time to time.

Storytelling is a time-honoured tradition that spans across all cultures. It serves as the fundamental means through which people, regardless of their backgrounds, initially acquire knowledge. This ancient art holds immense cultural value and plays a vital role in linking us to our past, forging our identities, and giving us a sense of purpose. It forms strong bonds within our families and communities.

An essential aspect of storytelling is its significant contribution to the emotional, cognitive, and linguistic development of children. For thousands of years, stories have been the vehicle for passing down essential knowledge and wisdom. A well-crafted narrative effortlessly ignites our curiosity, emotions, and imagination, enabling us to learn and hone skills with remarkable ease.

Storytelling is an invaluable gift to humanity, exemplified by Mike Lockett's adaptation of an eighth-century story in his work, *The Normal Storyteller:*

Many Centuries ago, a banquet was held in the palace of the Caliph of Bagdad to celebrate the birth of a new son. As was the custom, leaders and men of high station came from all across the land to present gifts and enjoy the feast provided by the Caliph. Each brought a costly gift, except for one wise sage named Mehelled Abi. Abi came to the feast empty-handed.

Each guest paraded in elegance before the Caliph to present his gift. Fine clothes, jewels and gold were given in abundance. As each gift was received with a smile from the Caliph, the givers turned and gave a frown as they watched the sage standing at the end of the line with no gift to give.

"He has no gift to give," they began to whisper among each other. "He has no gift to give the new prince."

At last, the sage came before the Caliph, and the room grew silent as all wished to hear what words he shared with their ruler. "Fine raiment, jewels and gold the prince has received, and may he be blessed by many more gifts as he grows to manhood. But I bring the most precious gift of all. I bring the gift of "stories". Laughter broke out in the room until the Caliph held up his hand to silence everyone and let the sage continue.

"From the time that your son is able to understand my words," said the sage, "I will come to the palace every day and tell him stories. The tales I tell him, both true and fanciful, will make him wise and righteous as he grows in wisdom through my words. When the day comes, my Caliph, that you can rule no longer, your son will sit above your people as their chief. He will be just and merciful, and all of Arabia will rejoice in his leadership."

Mehelled Abi kept his word. From the time the young prince could speak and understand words, the child, who was named Haroun-al-Rashid, was tutored daily by the wise old man, and told stories from all around the world - tales both true and fanciful. He learned of wise men and fools, of science and nature. He learned about all religions of the world and more. When the time came for him to rule as Caliph, he extended the borders of his empire from the Byzantine Empire in the West to China in the East. He ruled with wisdom like few other men possessed.

Stories, indeed, are a remarkable gift. They represent the most significant present received by Haroun-al-Rashid, a gift that left an enduring mark on his life. Abi's storytelling to Prince Haroun served multiple purposes: to entertain, to impart lessons from history, to provide insights into the complexities of the world, to unravel life's mysteries, and to instil the values essential for living a meaningful life.

In our early years, the first storytellers we encounter are our parents or caregivers, teachers, and the narratives woven into popular culture. It is during this impressionable period that the exposure to meaningful stories becomes pivotal, guiding us on the path of right living and the adoption of essential values. Stories have the profound ability to mould our lives and define our role in the world. Steve Jobs eloquently expressed this when he remarked, "The most powerful person in the world is the storyteller. The storyteller sets the vision, values, and agenda of an entire generation that is to come..."

The Story of Storytelling

Storytelling is an ancient art, as ancient as humanity itself. Humans started telling tales as they developed the capacity of speech more than fifty thousand years ago. Storytelling has been in use throughout time to heal minds, change perspectives, motivate, inspire and sooth. But who told the first story? When was it told? It is next to impossible to get answers to these questions. We can only make guesses as it is a retrospective reconstruction. Humans have been telling stories even before written history. The history of storytelling is shrouded in the mists of time, and we must fit loose pieces together to make sense of it. Storytelling is an ancient art and holds a revered place in most cultures. It connects us with our past and gives us identity and purpose. It connects us to our family and community. It dates to the origin of man's social life.

Visual Storytelling

For ages, storytelling was the main way of transmitting cultural knowledge and beliefs from one generation to the next. And they varied from teller to teller and listener to listener. It exists to entertain, to inform, and to promulgate cultural traditions and values. Over time, stories have undergone remarkable transformations in format, progressing from cave paintings to oral traditions, written records, and now digital storytelling. In the twenty first century, people make use of social platforms such as Facebook, Twitter, Instagram, and Podcast to tell their stories. In all cultures storytelling is central to its existence.

The ancient Chauvet cave paintings in Southern France, dating back approximately 36,000 years, provide a glimpse into this visual storytelling tradition, depicting animals, humans, and objects that appear to convey visual narratives. Similarly, Australia's Aboriginals communicated their stories through indigenous rock art millennia ago, offering early examples of pictorial storytelling. Egyptian hieroglyphics from around 3000 B.C. also reveal the roots of visual storytelling. This practice has since evolved from cave paintings to contemporary photography and film.

In the modern world, movies and TV series have emerged as some of the most prevalent forms of visual storytelling, carrying on this rich and timeless tradition of narrative through visuals.

Oral tradition

Long before the advent of writing, our ancestors relied on storytelling. They used spoken words and gestures to pass down stories, legends, history, and cultural knowledge in what we call oral tradition. This was the primary way they connected with each other and shared their heritage. Parents and grandparents would share these tales with their children, ensuring that family culture and traditions were passed down through the generations. It was likely that these stories were told in small groups, possibly around fires in caves, providing comfort and guidance in the face of fear and danger.

In ancient times, when people depended on hunting for their food, heroic stories of great hunters were shared around the hearth, inspiring young hunters to be courageous. As communities settled into more organized groups and clans, stories shifted to tales of heroes in times of war. Oral storytelling encompassed a wide range of narratives, including myths, legends, religious teachings, fairy tales, folk stories, poetry, chants, and songs. This practice helped people preserve their traditions, values, and cultural memories.

Storytelling gatherings were essential for building a sense of community and shared experiences. People came together to listen, learn, and connect, strengthening social bonds. There was an undeniable intimacy and connection between the storyteller and the audience. This is why oral storytelling is often described as dynamic, in contrast to written stories, which are more static. This distinction remains true even today.

In ancient times, people travelled more than we might imagine, and they carried their stories with them. These stories were shared along the road or by campfires. People back then perhaps had better listening skills, impressive memories, and were less concerned with precision. With each retelling, stories evolved, incorporating new elements to suit local interests, and becoming part of different cultures.

Storytellers held revered positions in society. They served as healers, leaders, spiritual guides, teachers, and entertainers. For a significant portion of history, storytellers played vital roles in their communities. The ability to tell stories effectively and memorably was a highly valuable skill for priests and rulers.

Before the widespread use of writing, storytelling served as the primary means of passing down knowledge to future generations. People had to rely on their memory to preserve this knowledge. While some stories were eventually written down, many were not. The advent of technology did not replace storytelling but rather revolutionized it. It made stories more accessible to people worldwide through mediums like radio and podcasts. Modern-day oral storytellers can now reach mass audiences, even the entire world simultaneously.

Written Storytelling

The first evidence of recorded storytelling is attributed to *The Epic of Gilgamesh* and later to Homer's epics *Iliad* and *Odyssey* and to *The Fables* of Aesop. These existed as oral stories long before they were ever written down. The invention of writing and much later, printing changed the course of storytelling, giving it shape and permanence. Literacy then was considered the monopoly of the elite in society. With the democratization of education, written storytelling became systematic, increasingly popular, and easily available to readers.

Digital Storytelling

The advent of digital storytelling has offered a wider platform to storytellers with innovative storytelling techniques. Thus, it has once again undergone a major transformation. The rise of radio, film, television, internet, has changed the way we tell stories. The internet gives us access to millions of stories at the click of a button. The use of social media has sharpened our storytelling skills. A digital story can combine the art of storytelling with multimedia elements such as images, music, and videos to create something compelling and engaging. Digital stories can be told through a variety of media, including photography, audio recordings, video recordings, and animation.

The Future of Storytelling

Will modern technologies take over storytelling or allow new ways of telling stories? Modern technologies like movies and TV shows already influence the way modern people tell tales. The mode of storytelling is constantly changing. Over time, it has become richer and more diverse. Technology has enabled new and exciting ways of telling stories. Today, stories are told through several different media and art forms. The chances are this trend will probably keep growing.

Many people consume content on mobile devices on social media. This has given rise to a new form of storytelling, short-form content which is perfect for mobile consumption. Human brain has evolved to process information quickly and efficiently, making this an attractive option for many people. Short-form content is appealing because it aligns with the fast-paced, attention-challenged nature of modern life, making it easier for creators to engage their audience quickly and effectively. However, it is important to note that the preference for short-form content does not mean long-form storytelling is obsolete. Both have their place and can be used strategically depending on the audience and the intended message or impact.

Intrinsically, people are storytellers. But man's desire to tell stories and listen to them will continue because we crave for human connection. From cave paintings to virtual reality, storytelling has endured for thousands of years, and will continue to shape our world for generations to come. As a species, storiescontinue to captivate us. With the passage of time, we have become exemplary storytellers. It is only the medium of storytelling that has changed, not our interest in telling stories or listening to them.

Storytelling with Children

"If you want your children to be smart, tell them stories. If you want them to be brilliant, tell them more stories." Albert Einstein

Storytelling is the best way to communicate ideas to children. Storytelling arouses imagination, curiosity and enhances perceptual growth.

There is an anecdote about Albert Einstein's love of fairy tales. Once a mother who was full of admiration for him wanted to make her son a great scientist like the famous physicist. She sought his advice on how to raise her son to become a successful scientist. She wanted to know what kind of books she should read to her son.

To her great surprise, Einstein replied, "Fairy tales."

"What else should I read to him after that?" the mother asked.

"More fairy tales," Einstein replied.

She expected a more serious answer "And then?"

"Even more fairy tales," the great scientist added.

Einstein placed imaginative thinking on a more important level than knowledge. He said, "Great inventions require an imaginative mind." Einstein considered fairy tales important because they stimulate imagination. He said, "Imagination is more important than knowledge. For knowledge is limited whereas imagination embraces the entire world, stimulating progress, giving birth to evolution."

Telling stories to children is a multifaceted activity that contributes significantly to their growth and development. It is a particularly helpful form of therapy with children because stories are already a normal and interesting part of their life.

Stories instruct children about the world in a safe, caring environment. It stimulates their cognitive abilities, nurtures their emotional well-being, and fosters connections with caregivers and the wider world, all while being a delightful and enjoyable experience.

There are many reasons why storytelling is important for children.

Language Development

Storytelling exposes children to a rich and varied vocabulary. Hearing stories helps them learn unfamiliar words, sentence structures, language patterns, creative thinking, concept formation and listening skills which in turn aids in their communication skills. Exposure to stories and storytelling helps prepare children for reading and writing. They begin to recognize letters, words, and the structure of written language, setting the stage for literacy development. It is easy to teach the meanings of the words as kids learn faster from the context of story.

Let's consider an anecdote to illustrate the impact of storytelling on language development:

Ezra's Language Journey

Once upon a time, in a small town, there lived a curious little boy named Ezra. His parents were enthusiastic storytellers, and they made it a nightly ritual to share captivating tales before bedtime. One of Ezra's favourite stories was about a magical forest where animals could talk.

As Ezra listened to these stories, his eyes widened with wonder. One day, as he played in his backyard, he spotted a peculiar-looking bird. Remembering a story about talking animals, he approached the bird with excitement.

To his surprise, the bird chirped in a language only Ezra could understand. Over the weeks, Ezra continued interacting with different animals, each with its own unique way of communicating. Through these encounters, he not only expanded his vocabulary with words like "chirp," "whisper," and "giggle" but also learned about sentence structures and language patterns.

When Ezra started school, his early exposure to storytelling proved invaluable. He easily recognized letters and words thanks to the varied vocabulary he had encountered in the magical forest stories. The structured narratives had laid the foundation for his understanding of written language.

Ezra's teacher noticed his advanced language skills. One day she asked the class to write a short story. Ezra enthusiastically wrote down his own tale, incorporating the lessons he had learned from the stories at home. His classmates were amazed by his creative thinking and expressive language.

Ezra's journey highlights how exposure to storytelling enriched his vocabulary, nurtured his creative thinking, and played a pivotal role in his early language and literacy development.

Imagination and Creativity

Storytelling can inspire children to use their imagination in play. They may re-enact scenes from the story or create their own stories with characters who mirror their own experiences. This type of imaginative play can help them process and work through emotions. Stories transport children to different worlds, settings, and scenarios. This sparks their imagination and encourages creative thinking. Children can visualize characters and places in their minds, which is a crucial skill for problem-solving and creativity. When children listen to stories, they respond by creating images of the characters and places described by the words. When a child listens to a story the absence of visual cues helps them imagine the story and process the tale.

Let me elaborate this with a simple story.

The Enchanted Seeds of Imagination

Once upon a time, in a small village, there lived a curious young boy named Vishak. Every evening, as the sun dipped below the horizon, the village storyteller, old Bashir, would gather the children around a crackling fire.

One evening, Bashir spun a tale of animals in a magical woodland that held secret meetings under the moonlit canopy. Vishak's eyes widened with wonder as he imagined shimmering fireflies lighting up the enchanted glade. In his mind, he could hear the wise old owl hooting advice to the mischievous squirrel, who was always up to playful tricks.

Inspired by the story, Vishak spent the following days re-enacting the forest scenes with his friends. Together, they transformed the village outskirts into their magical woodland, each child taking on the role of a different creature. Vishak, with a crown of wildflowers, became the guardian of the secret meetings.

As they played, Vishak and his friends discovered the joy of collaboration and problem-solving. They invented new twists to the story, creating unexpected challenges for their characters to overcome. In this imaginary world, Vishak found a space to express and understand his emotions, just like the characters in old Bashir's tales.

The village storyteller smiled as he watched the children bring his words to life. Little did they know, in the dance of flickering firelight and youthful laughter, they were not just playing - they were cultivating the seeds of imagination and creativity that would blossom throughout their lives.

Cognitive Development

Incorporating storytelling into a child's daily routine, whether through reading books, telling stories aloud, or engaging in interactive storytelling, can have a profound impact on their cognitive development. It fosters a love for learning and sets the stage for success in various academic and intellectual pursuits. Stories often have complex plots and characters, which challenge children's cognitive abilities. They learn to follow sequences, make connections between events, and develop their critical thinking skills as they analyse the story. Stories from diverse cultures and traditions introduce children to diversity and broaden their understanding of the world. Stories often feature characters from various cultural backgrounds. By encountering characters with different ethnicities, languages, and customs, children can learn about the diversity of the world. This promotes tolerance and cultural awareness.

Let me share a story to show how storytelling fosters cognitive development.

The Enchanted Journey

Once upon a time, there was a curious young turtle named Sumit. His mother, Tina, loved telling bedtime stories. One night, she shared a tale about Sumit's distant relative, an adventurous turtle named Akash.

As Sumit listened, he practised following the sequence of events. "First, Akash set out on a journey. Then, he encountered a tricky river he had to cross," Tina narrated. Sumit's mind worked to piece together the story's order.

Tina asked, "What do you think Akash should do next?" This simple question engaged Sumit's critical thinking. He pondered different options, considering Akash's choices and their consequences.

The story introduced Sumit to new words and phrases, expanding his vocabulary. Tina described Akash's encounters with animals of various backgrounds, subtly promoting cultural awareness in Sumit.

With vivid descriptions, Sumit's imagination soared. He envisioned Akash's journey, the challenges he faced, and the friends he made. This imaginative exercise stimulated Sumit's creativity.

As Tina concluded the story, her son drifted into sleep, the cognitive seeds planted. Through storytelling, his mind had danced with sequences, critical thoughts, new words, diverse perspectives, and imaginative scenarios – a beautiful symphony fostering cognitive development.

Morals and Values

Many stories contain moral lessons and ethical dilemmas that encourage children to think about right and wrong. They conclude with resolutions that align with positive values, such as fairness, justice, or kindness prevailing. These endings leave children with a sense of hope and the idea that living by certain values can lead to positive outcomes. The values such as the importance of sharing, the value of time, compassion for others, courage honesty etc. learned from stories often stay with children throughout their lives. These values can influence their decision-making, behaviour, and relationships as they grow and mature.

Stories such as the following one instil values in us.

Echoes of Empathy

In a small town, there were two friends, Arya and Neha, who loved spending their weekends at the local park. One day, they noticed an elderly woman, Mrs. Kapoor, struggling to carry her groceries home. Arya and Neha quickly rushed to help her, offering to carry the bags.

Grateful for their assistance, Mrs. Kapoor invited them into her home for a cup of tea. As they sat together, Mrs. Kapoor shared stories from her youth and the challenges she faced as a single parent. Arya and Neha listened intently, realizing the importance of empathy and understanding.

Over the next few weeks, the two friends continued to visit Mrs. Kapoor, helping with household chores and spending time chatting. They learned about the value of time, the importance of compassion for others, and the strength that comes from facing life's challenges with courage.

One day, Mrs. Kapoor fell ill, and Arya and Neha took turns caring for her. They realized the impact of their friendship on Mrs. Kapoor's well-being and the positive outcomes that arise when people extend a helping hand to those in need.

As years passed, Arya and Neha's bond with Mrs. Kapoor deepened. The lessons of kindness, compassion, and the value of time they learned from their experiences became guiding principles in their lives. The simple act of helping a neighbour had not only strengthened their friendship but had also instilled in them a lasting commitment to positive values.

Bonding and Connection

Sharing stories can be a bonding experience between adults and children. Children often look forward to their bedtime as there is an incredible story session that lies in the wait. This entire session of storytelling can be rewarding for the child as well as the parents. It provides an opportunity for quality time together and strengthens the parent-child relationship. Storytelling provides an opportunity for quality one-on-one time. Whether it is a parent reading to a child before bedtime or a teacher reading to a classroom, the act of storytelling allows for focused and uninterrupted interaction.

Attention and Listening Skills

Storytelling is a powerful tool that not only entertains but also encourages and enhances the crucial listening skills of children. In classrooms, where the tendency to talk often outweighs the inclination to listen, cultivating a habit of attentive listening can be challenging. However, introducing engaging activities that promote active listening can make a significant difference.

Consider the example of Miss Mia, a creative teacher who recognized the importance of fostering attention and listening skills in her students. She introduced a game called "The Whisper Circle". She gathered the students in a circle and whispered a short sentence to the first child. The task was to pass the message by whispering it to the next person until it reached the last student.

The sentence was simple: "The happy cat danced in the sunlight." As the message travelled around the circle, each child concentrated on listening carefully to pass it on accurately.The teacher then asked the last student to share what he heard.

To everyone's amusement, the final sentence was, "The snappy hat pranced in the moonlight." The children burst into laughter, realizing how the message had transformed. Miss Mia seized the opportunity to discuss the importance of attention and listening skills. This simple yet effective activity emphasized the significance of attentive listening. The children not only found the game enjoyable but also learned the value of patience and focused attention. She explained that just like in the game, paying close attention ensures clear communication and prevents misunderstandings.

When a habit of listening to stories is inculcated in them, children become better listeners. It provides them with essential training to focus on and understand more, encouraging them to refrain from constant talking. Through storytelling, children learn how to pay attention and listen actively to the person talking. "The Whisper Circle'" game served as a practical example of how such interactive listening activities contribute to active engagement and skill development.

Moreover, children's love for listening to the same story repeatedly can be seen as a reinforcement of their listening skills. Children often notice added details or gain a deeper understanding of the story with each reading or listening session. Repeated exposure to stories and the subsequent triggering of mental images stimulates appropriate neural development in the brain. This is the reason that children insist on hearing the same story repeatedly.

Miss Mia's students, inspired by "The Whisper Circle" experience, not only insisted on revisiting the same story but also demonstrated improved listening skills, gaining a deeper understanding with each repetition.

Memory Enhancement

One of the great benefits of storytelling is that it boosts memory, improves listening skills and concentration. Remembering the details of a story and recalling events or characters later helps improve memory and retention skills. Encouraging children to retell the story to others or extend the story through their own creative activities reinforces memory. This requires recalling and summarizing key details. Effective storytelling techniques can make learning enjoyable and memorable, helping children retain and recall information more effectively. Stories help us remember and integrate what we learn. It is said recall is twenty-two times stronger when in story format rather than just data. Stories are effective educational tools because listeners become engaged and, therefore, remember. Effective stories cause us to feel emotions. Emotions heighten our ability to memorize experiences and thus help improve information processing. Stories make it easier for our brains to store data for later retrieval. Emotions are a signal to the brain that whatever we are experiencing is important.

Stress Reduction

Engaging in storytelling can be a comforting and soothing experience for children. It can help reduce stress and anxiety, particularly when shared before bedtime. Regular story time routines create a sense of security for children. Knowing that they will have this comforting and predictable experience can reduce anxiety and stress. Stories often depict characters using diverse coping strategies to navigate challenges and stress. Kids can gain insight into effective stress management and emotional regulation by observing how characters handle similar situations. Valuable coping strategies, like taking deep breaths or seeking support from trusted friends, can be gleaned from these narratives.

After the narration, the storyteller can create a safe environment for kids to talk about their feelings. The narrator may ask questions like "Do you understand how the character felt?" or "Did something like this ever happen to you?" This helps them share what they think and feel, which is good for their emotions.

Entertainment and Enjoyment

Most importantly, storytelling is a source of joy and entertainment for children. It cultivates a love for reading and learning that can last a lifetime. Depending on the age of the children, you can incorporate visual aids like pictures, puppets, or props to enhance the storytelling experience. Visuals can help children better understand the story and make it more enjoyable. After telling a story, encourage children to create their own stories or drawings related to the story they heard. This fosters creativity and helps them connect with the story on a deeper level. Storytelling strengthens your bond with your children through shared laughter, joyful and positive memories. Moreover, it helps to promote a positive attitude in your children towards life from a tender age that will last a lifetime. Be creative while you read the story to your child. Usually, a personality or character of the story catches a child's interest the most. So, use different voices for each character or act out various parts from the story, to bring life to those characters. Acting out parts from the story will make story sessions fun and memorable for your kids. Place books or pictures in front of your kids, inviting them to turn the pages or trace the words as you read.

Identification and Expression of Emotions

Storytelling can be a powerful tool for emotional healing in children. Stories frequently involve characters experiencing a wide range of emotions, allowing children to learn about and understand their feelings better. They can relate to characters' emotions, which can help them process their own. Stories often feature characters who experience a wide range of emotions, just like children do. By listening to or reading these stories, children can identify with the characters and begin to understand their own feelings better. They may see characters expressing sadness, fear, anger, or joy, which can help them feel less alone in their emotions.When the storyteller discusses the emotions of the characters in the story, he/she can also validate the child's feelings. For instance, he/she might say, "Just like how the character in the story felt sad when she couldn't find her teddy bear, it's okay for you to feel sad too." This validation helps the child feel understood and accepted.

We feel empathy with the characters we encounter in stories. Hearing about characters who face challenges or difficult emotions can promote empathy in children. They learn to put themselves in someone else's shoes and imagine how others might be feeling. This fosters a sense of compassion and understanding towards others. This can be particularly helpful in resolving conflicts and building positive relationships.

Emotional healing is often a gradual process. Through storytelling, children can experience a sense of emotional relief and release, forging connections with characters and their journeys. Little Emma, facing her own emotional struggles, found solace in such tales.

Emma's Healing Connection

In a quiet room, she listened to a story about a character confronting loneliness, a theme that resonated deeply with her. As the tale unfolded, Emma felt a sense of connection and understanding, realizing that the character's journey mirrored her own. This experience allowed her to find a path to healing, providing a gradual but comforting release from her inner struggles, as consistent exposure to these narratives became a guiding light for her emotional journey.

Talking with the child after a story helps them feel better. Storytelling is just one way to help them feel good emotionally. Sometimes, it is important to get help from professionals too.

Behavioural Disorders in Children

Behavioural disorders in children refer to a wide range of mental health conditions that affect a child's behaviour, emotions, and social interactions. These disorders can manifest in many ways and can have a significant impact on a child's development, relationships, and overall well-being. For children with behavioural disorders like ADHD, oppositional defiant disorder, eating disorders, anxiety disorders, storytelling can be used to teach social skills, self-regulation, and problem-solving. Stories featuring relatable characters facing similar challenges can serve as models for behaviour change.

Breaking Barriers

Once there lived a spirited young boy named Ehan who grappled with Oppositional Defiant Disorder (ODD). Ehan's days were often filled with moments of frustration, and his relationships with others were strained.

One day, his teacher, Mrs. Safia, introduced a unique project to the class. It was a story-sharing initiative where each student could create and share a tale based on their own experiences. Ehan, intrigued by the idea, decided to craft a story reflecting his challenges and triumphs.

In Ehan's story, the main character, Hanif, exhibited similar oppositional behaviours but embarked on a journey of self-discovery. Through the narrative, Ehan explored the emotions and struggles that often consumed him. The character of Hanif, much like Ehan, encountered patient friends and supportive adults who guided him toward positive choices.

As Ehan shared his story with the class, his classmates gained insight into the daily battles he faced. Rather than judgment, they offered understanding and encouragement. The power of storytelling became a bridge between him and his peers, fostering empathy and breaking down the barriers created by ODD.

Over time, as Ehan continued to share his experiences through stories, the classroom transformed into a supportive community. The positive impact of this storytelling initiative extended beyond understanding ODD; it created an environment where everyone, regardless of their challenges, felt heard, accepted, and capable of positive change.

Building Resilience

Stories often depict characters facing adversity and finding solutions. When children witness characters overcoming challenges, they can develop a sense of resilience and belief in their ability to tackle their own difficulties. Stories often portray characters who overcome obstacles or deal with their emotions in healthy ways. Stories can play a significant role in helping children develop resilience, which is the ability to bounce back from adversity, cope with challenges, and adapt to demanding situations. Many stories involve characters who use creative problem-solving strategies to overcome obstacles. Children can learn from these examples and develop their own problem-solving skills, which are crucial for resilience.

Healing Through Symbolism

In some cases, storytelling can use symbolism to address difficult emotions or experiences indirectly. Metaphors and allegories can be used to explore complex topics in a way that feels safer for the child. Symbolism provides a safe and imaginative space for children to explore difficult emotions or experiences without directly confronting them. This can be especially helpful for children who may be reluctant or unable to express their feelings openly.

Allow me to share a story that illustrates the power of symbolism in healing:

Maya's Butterfly

Once there lived a young girl named Maya who was going through a difficult time coping with her parents' separation. Her teacher, Mrs. Sharma, noticed the girl's withdrawn behaviour and decided to help through the power of storytelling.

One day, Mrs. Sharma began telling a story about a butterfly that faced challenges in its cocoon before emerging transformed and beautiful. Maya listened intently, connecting with the butterfly's journey as a metaphor for change and growth.

As the story unfolded, the teacher delicately wove in elements that mirrored Maya's feelings of confusion and sadness. The cocoon represented the challenges of her parents' separation, and the butterfly symbolized the potential for beauty and resilience within her.

After hearing the story, the girl felt a sense of relief. Mrs. Sharma created a safe space for Maya to indirectly explore her emotions through the symbolism of the butterfly. Over time, Maya began to open up, sharing her thoughts and feelings more openly.

The symbolic story became a powerful therapeutic tool for the student, helping her navigate the complex emotions surrounding her family situation. Mrs. Sharma's thoughtful use of symbolism provided Maya with a way to understand and accept her feelings, offering hope for positive change and growth.

This small story illustrates how symbolism in storytelling can be a gentle yet effective method for addressing difficult emotions, especially in the context of a child dealing with personal challenges.

Stories with symbolic elements can serve as metaphors for the process of change and growth. Children can identify with characters who undergo transformation and find hope in the idea that change is possible. When using symbolism in stories for healing, it is essential to be sensitive to each child's unique needs and experiences. It can be helpful to work with trained therapists or counsellors who specialize in using storytelling and symbolism as therapeutic tools, especially for children dealing with trauma or deep emotional issues. Symbolic stories should be selected carefully to ensure that they resonate with the child's age, developmental stage, and specific challenges.

Children are often not able to process and verbalize their traumatic and uncomfortable feelings. Successful intervention is possible using metaphor in storytelling, and it avoids confrontation and interpretations. Research has proved that storytelling brings about changes in children who are abused and traumatized. Stories help children to accept family break ups and bereavements. It helps them to cope with feelings. Stories that are affirming and that contain positive messages are told or read to children. This therapeutic technique can be used not only by professionals in the field, but also by parents, teachers, and caretakers of children.

Teachers often use stories to teach complex concepts to their students. By bringing history or science to life with a well-told story, teachers can help their students make connections and deepen their understanding of the material. With imagination, with the ability to understand symbols, create solutions, and find meaning in ideas, young people are more capable of mastering language, writing, mathematics, and other learnings that are grounded in the use of symbols. Children need interesting things to talk about; things that capture their curiosity and fire their imagination.

The Healing Art of Storytelling

"All stories have therapeutic or healing potential. If a story makes people laugh, the laughter can be healing. If a story makes them cry, this can be healing too. Folk and fairy tales, through their universal themes and resolutions, have healing possibilities. They can offer hope and courage to face the trials of life and help the listener find ways to move forward." (Perrow)

Storytelling is the most ancient form of psychotherapy. It can make people feel better, calm them down, make them think differently, and teach them things even without their realizing it. When someone tells a story, it can teach and heal at the same time. It is like taking a journey in your mind instead of getting a lecture or someone telling you directly about a problem. Storytelling also helps us feel closer to other people in real life. The special thing about storytelling is that it can touch all parts of our lives. Stories have a unique power to bring people together and make them feel better, whether they are dealing with emotional or spiritual struggles. Throughout history, storytelling has been used to help people's minds, change how they see things, motivate them, inspire them, and make them feel better. A story, with its power to unite and connect people, can break down walls between individuals and heal their wounds.

Stories can be told for therapeutic purposes - to help people with their problems, whether they are physical, psychological, or emotional. They can also be used to learn new skills. Stories are a great tool for therapy, in education, within families, for children, and in everyday life. Narrative therapy is a type of therapy that focuses on using stories. Therapeutic storytelling is when you tell stories to understand yourself or what is happening in your life better. Stories do not just help the person listening; they also help the person telling the story feel better.

We have been using storytelling to help others feel better for a long time, no matter how old they are. This can be helpful in psychotherapy, counselling, and when trying to help yourself. It is important to note that the effectiveness of therapeutic storytelling may vary from person to person. Some people really like it, while others might like different therapeutic approaches. The most important thing is to use storytelling in a way that works best for each person. It often works best when a therapist or counselor, who knows how to do it well, is involved.

Why does therapeutic storytelling work so well?

1. We all do not like change much. When we listen to a story, we do not worry too much about making sense of everything. We put our logical thinking on hold for a bit.
2. Stories can make us feel strong emotions. When we get caught up in a story, we might forget about being too logical. Emotions can be so strong that they push our logical thinking aside, and we start feeling what the characters in the story feel.
3. Stories let us escape from real life. Sometimes, we just want a break from our everyday routines and problems. When we dive into a story, we are looking for an adventure, and we are fine with letting go of logic for a while.
4. Some cultures have strong storytelling traditions. In those places, people expect stories to have all sorts of magical or illogical stuff, and they wholeheartedly approve of it. Culture can affect how willing we are to go along with stories that do not make complete sense.
5. Stories are supposed to be fun and captivating. When we are enjoying a delightful story, we are willing to go along with it, even if it does not make perfect sense. It is all about getting lost in the story's world.

Therapists often use storytelling techniques such as narrative therapy, bibliotherapy, and journaling to help people process trauma, grief, addiction, or other mental health issues. Storytelling is a powerful and healing way to cope with trauma and stress. It can help you express, reframe, and connect with your experiences, feelings, and insights. It can also help you develop your oral communication skills, which are essential for personal and professional success.

Here are some reasons why therapeutic storytelling can be effective with adults:

Emotional Healing Through Storytelling

Healing your emotions through storytelling is a powerful way to express and process your feelings creatively and safely. Sharing your personal stories, especially for those who have faced trauma or difficult experiences, can provide comfort and healing. It helps you realize that your emotions are valid, and you are not alone in your struggles.

Storytelling allows us to communicate our thoughts and emotions effectively, making it easier to explore our feelings safely. When we share and listen to stories, we gain insight into others' experiences, fostering connection and the opportunity to discuss our emotions.

Telling stories not only facilitates expressing thoughts and emotions but also helps build trust among individuals, an essential component of healthy relationships. Storytelling serves as a constructive means of resolving conflicts and misunderstandings by allowing individuals to convey their emotions and perspectives.

Additionally, sharing stories about our experiences, culture, or community enhances our sense of belonging and self-understanding. Engaging in storytelling addresses various emotional and social needs, promoting empathy, trust, and connection, which are vital aspects of human communication and relationships. Understanding others' perspectives and emotions fosters a sense of belonging and supports emotional healing.

Now, consider the broader perspective on emotional healing through storytelling:

Connecting Through Shared Narratives

In a small support group for trauma survivors, Daisy hesitated before sharing her story.

As she began to speak, the room fell silent, enveloped in the poignant rawness of her emotions. As Daisy recounted her experiences, others in the group nodded in understanding, some even sharing similar feelings.

This simple act of storytelling transformed the group into a community of shared struggles and resilience. The exchange of emotions in words not only provided relief for Sarah but also created connections among the group members.

Metaphor and Symbolism

Stories often contain metaphors and symbolism that can resonate with individuals on a deeper level. The use of metaphor tends to bypass psychological resistance. Metaphorical stories can serve as a mirror for clients to better understand their own feelings and behaviours. Such stories often include elements of transformation or growth, offering clients hope and inspiration for positive change. These stories can convey that change is possible and that they can overcome their challenges and emphasize the idea of resilience and adaptability in the face of adversity. People can learn to see themselves as the heroes of their own stories who have the power to overcome obstacles. Metaphors can represent inner conflicts and dilemmas, allowing them to externalize and negotiate these conflicts within the context of the story. This can help them gain insights into their conflicting thoughts and emotions, paving the way for profound self-discovery as we can observe in the following example.

Breaking Through Shadows

In a counselling session, Nirmal grappled with the weight of past mistakes, guarded by layers of self-blame. The counsellor, sensing the psychological resistance, wove a metaphorical tale of a resilient tree weathering storms, shedding old leaves to make way for new growth. As the story unfolded, Nirmal saw parallels to his own journey, realizing the possibility of shedding guilt and fostering personal growth. The metaphor, like a powerful remedy, penetrated through resistance, offering a symbolic reflection that opened avenues for self-discovery.

Self-Esteem and Self-Confidence

Stories that focus on characters developing self-esteem and self-confidence can inspire individuals to work on their self-worth and belief in their abilities. Stories often include characters who achieve success or personal growth. Experiencing these stories can boost self-confidence by showing that success is attainable through effort and determination. The following story is an instance in point.

Brushstrokes of Confidence

Once there was a young artist named Kamala who loved painting but struggled with self-doubt. Despite her talent, Kamala often hesitated to share her artwork, fearing judgment from others. One day, she stumbled upon an old art studio where a wise mentor, Mr. Husain, noticed her hesitation.

Mr. Husain shared tales of his own journey, of once doubting his abilities. Inspired by his stories, Kamala began to experiment with bolder strokes and vibrant colours, slowly regaining confidence in her art. As she embraced the process of storytelling through her paintings, Kamala discovered a newfound sense of self-expression and self-worth. The once hesitant artist transformed into a confident painter, illustrating the transformative power of stories in reigniting one's self-confidence.

Anxiety and Stress Management

Engaging in creative storytelling can be a relaxing and stress-reducing activity for adults. It provides an opportunity to escape from the stresses of daily life and focus on creative expression. Therapeutic storytelling can introduce positive and calming imagery. Imagining serene landscapes, peaceful scenarios, or soothing environments can help reduce stress and induce relaxation. Stories that focus on relaxation, mindfulness, or overcoming stressful situations can help individuals manage anxiety and stress. Narratives that offer coping strategies and depict characters successfully managing stressors can be particularly helpful. One of the many potential health benefits of storytelling is its relaxing effect. The act of sharing stories can help reduce stress levels and promote mental clarity, which can have a tremendous impact on our physical health.

The following anecdote offers a simple coping mechanism for anxiety and stress.

Pebbles of Tranquillity

In the little town of Serenity Springs, lived a young woman named Pranika. Despite her outward calm, she often found herself overwhelmed by stress and anxiety. One day, her wise neighbour, Vanitha, noticed Pranika's struggles and shared a simple yet effective coping strategy.

Vanitha handed Pranika a small jar filled with colourful pebbles. Each pebble represented a different coping mechanism. "Whenever you feel the weight of stress, take a moment," Vanitha advised. "Hold one pebble, breathe deeply, and focus on the corresponding strategy."

Pranika began using the jar whenever stress struck. The blue pebble reminded her to take slow, deep breaths. The green one prompted her to visualize a serene meadow. The red, a symbol of warmth, encouraged reaching out to a friend or loved one.

As the days passed, Pranika found solace in the pebbles. They became her allies in moments of stress, guiding her towards calm and clarity. Through this simple narrative, she discovered that amidst life's chaos, small coping strategies could pave the way to peace and tranquillity.

Anger Management

Stories that highlight characters dealing with anger and finding healthy ways to express and manage their emotions can be effective in teaching anger management skills to both children and adults. Therapeutic storytelling can prompt individuals to reflect on their own anger triggers, patterns, and the underlying causes of their anger. Through discussions or journalling, they can connect the story's themes to their own experiences and gain insight into their anger. Therapeutic storytelling can inspire individuals to set goals for their anger management journey. They may identify specific changes they want to make in their behaviour, relationships, or emotional responses and work toward these goals with guidance from the stories. Over time, exposure to therapeutic stories can contribute to the development of emotional resilience. Individuals may become better equipped to manage anger in challenging situations and adapt healthier coping mechanisms.

The following story highlights the transformative power of self-awareness and creative therapeutic techniques in the realm of anger management.

The Anger Tree

Once there lived a woman named Maria. She had a demanding job, and the stress of her daily life often left her feeling overwhelmed. However, her biggest challenge was managing her quick temper.

One day, after a particularly stressful meeting at work, Maria decided it was time to tackle her anger issues. She sought the help of a therapist, Dr Sagir, who suggested an unconventional approach.

He introduced Maria to the concept of the "Anger Tree." They discussed identifying the roots of her anger, deep-seated frustrations, and stressors. The trunk represented the build-up of tension, and the branches symbolized the outward expressions of her anger.

Maria started a journal to track her emotions, noting down situations that triggered her anger and the corresponding physical and emotional responses. As Maria reflected on her entries, she noticed patterns and began to understand the root causes of her anger.

Dr Sagir then guided Maria in visualizing her "Anger Tree" during moments of frustration. As she imagined the tree, she consciously worked to address the roots, focusing on stress reduction strategies and coping mechanisms. She also practised deep-breathing exercises to calm the trunk and redirect the energy that would have fuelled her anger into more constructive outlets.

Over time, Maria's "Anger Tree" evolved. The roots became less entangled, the trunk less rigid, and the branches less volatile. She found herself responding to challenging situations with newfound composure.

Through this journey of self-discovery and visualization, Maria learned to manage her anger effectively. The "Anger Tree" became a powerful tool, helping her navigate the complexities of her emotions and fostering a sense of control over her reactions.

Trauma and PTSD

Narrative therapy is often used to address trauma and post-traumatic stress disorder (PTSD). Personal storytelling in a therapeutic context can help individuals process and heal from traumatic experiences. It can help individuals integrate fragmented memories of the trauma into a coherent narrative. This process can reduce the intrusiveness of flashbacks and nightmares associated with PTSD. Therapeutic storytelling can incorporate self-compassion themes, helping individuals be kinder to themselves as they navigate the healing process. It can counteract self-blame and guilt associated with trauma.

Individuals may carry the remnants of their past experiences as exemplified by this story:

A Tale of Tails

Once a fish called Veer and an old frog, Harit lived together harmoniously as neighbours and friends. One day it was rumoured that the ruler of the sea, Varuna, had ordered all creatures with tails beheaded. Veer was terrified and swam to his old friend for comfort. But when the fish reached Harit's place, he was surprised to find the old frog shivering with fear.

"Harit, what has happened to you? Why are you so frightened? asked Veer.

"Haven't you heard that Varuna is going to kill all tailed creatures?" asked Harit.

"My dear friend, you don't even have a tail. What are you worried about?" Veer asked.

"I too had a tail when I was a young tadpole," replied Harit continuing to shake like a leaf.

We all have our own childhood "tails" which haunt us from time to time. But since these tails are in the past, why should our present be preoccupied by them?

In the journey of healing from trauma, individuals often carry the echoes of their past experiences, much like the old frog remembered its tadpole tail. The therapeutic power of storytelling lies in weaving these fragments into a coherent narrative, allowing individuals to make sense of their experiences. Just as the old frog's past tail did not define its present, personal storytelling can assist in reducing the intrusiveness of flashbacks and nightmares associated with PTSD. Moreover, incorporating self-compassion themes, as seen in the tale of the frog and the fish, becomes a guiding light, helping individuals be kinder to themselves as they navigate the intricate path of healing and countering self-blame and guilt associated with trauma.

Empowerment

Personal storytelling empowers us by providing narrative control and fostering positive reframing. Stories, serving as mirrors, enhance self-awareness, comfort, and resilience. Therapeutic storytelling challenges negative thought patterns and aids perspective reframing, facilitating self-reflection. This process guides individuals on a transformative journey toward a positive future by identifying values, strengths, and growth areas.

Should The Therapeutic Story Be Discussed After It Is Told?

"Just tell the story and let the listener figure it out. Altogether too many preachers and speakers insult the intelligence of

their listeners by telling a story and then promptly draining it of all meaning by a lengthy analysis.... This literary form is both concise and direct." (White)

There should be no discussion about what the story meant; trust their intuitive ways to understand the meaning. If the child is contemplating the meaning of the story or appears reluctant to discuss it, that response needs to be respected. Sometimes it may be better to allow listeners to engage in their own search for meaning. Interrupting this process might hinder their useful time of processing. If the child wants to talk about the story, not doing so seems to devalue his or her needs. The intent behind telling the story may not be the message the listener derives from it. Healing stories are likely to have their greatest impact when you assist listeners in seeking and finding their meaning - and that may be by quietly letting them discover it for themselves.

The Power of Personal Storytelling

Expressing your story through various forms like writing, art, or spoken word can be profoundly therapeutic. These forms of expression enable you to externalize your emotions and experiences, making them more manageable and facilitating the processing of your thoughts and feelings.

In Girish Karnad's *Nagamandala,* one encounters a theatrical work delving into diverse themes, among them the potency of storytelling and the societal role of women. Karnad employs storytelling as a channel through which characters articulate their aspirations, apprehensions, and sentiments. The play highlights the importance of sharing one's individual narrative.

New flame: My mistress, the old woman, knows a story and a song. But all these years she has kept them to herself, never told the story, nor sung the song. So, the story and the song were being choked, imprisoned inside her. This afternoon the old woman took her usual nap after lunch and started snoring. The moment her mouth opened, the story and the song jumped out and hid in the attic. At night, when the old man had gone to sleep, the story took the form of a young woman, and the song became a sari. The young woman wrapped herself in the sari and stepped out, just as the old lady was coming in. Thus, the story and the song created a feud in the family and were revenged on the old woman.

Flame 1: so, if you try to gag one story, another happens.

... Story: You cannot just listen to the story and leave it at that. You must tell it again to someone else." (Karnad)

In *Nagamandala*, Karnad uses storytelling to explore themes like speaking one's mind, breaking free from society's rules, tackling taboos, and how stories can change us. Sharing your own story or singing your own song shows how important it is to express yourself and be empowered, even when society tries to hold you back.

We often feel unsure about sharing our own story because it means telling others about our personal thoughts and feelings. This can make us feel vulnerable and worried about what others might think or say. When we share our story, it can bring back the feelings we had during those times, which can be hard. And how others react to our story can also affect our emotions, whether they show understanding or don't seem to care.

Vulnerability is not weakness. It's honesty and trust in their pure forms. It's baring your soul and letting others see your darkness. It's trust in humanity that you won't get even more hurt, but someone might lift you up or stand by you until you lift yourself up.

I write about my life for a few reasons. Not to make money off my pain. Not to ask for sympathy. Not to subtly complain-brag about my progress.

I tell my story to heal myself and to help others. Because writing has a healing capacity. (Fontaine)

Sharing stories can be a powerful and helpful way for people to feel better and become better. Personal stories can be about your own experiences, worries, happiness, or dreams. They can also be a way to have fun and feel inspired.

Writing down your thoughts and feelings can be helpful for your mind and body, especially if you have experienced something difficult like PTSD. It can make you feel less anxious and angry and help you relax physically and focus better.

This kind of therapy also involves talking about your own stories, whether they are made up or real, and writing your own stories. Sharing our own stories helps us make sense of the world and unwind from stress. Writing stories can help us find meaning in tough times and fix things that got messed up because of illness. It can also give us the strength to move on with our lives.

If you feel happy with it, you can share your stories with someone you trust or a small group of people. Talking about your experiences can make you feel better, and when others listen, it shows that they understand what you are going through. Sharing stories is a strong way to express yourself and learn more about yourself.

Everyone can talk about their own stories because everyone has one. Sharing our stories is important for healing. When we write about our experiences, it is a way of being kind to ourselves. So, let us start writing the stories of our lives, whether we share them with others or not. It is all about using storytelling to reconnect with who we are.

"Every time you tell your story and someone else who cares bears witness to it, you turn off the body's stress responses, flipping off toxic stress hormones like cortisol and epinephrine and flipping on relaxation responses that release healing hormones like oxytocin, dopamine, nitric oxide, and endorphins. Not only does this turn on the body's innate self-repair mechanisms and function as preventative medicine - or treatment if you are sick. It also relaxes your nervous system and helps heal your mind of depression, anxiety, fear, anger, and feelings of disconnection." (Rankin)

Telling your own stories can help you understand your life better. It is a way to deal with tough feelings, like sadness, anger, or shame. Sharing your story or hearing others' stories can make you feel better and help you let go of your bad feelings. This is especially helpful if you've been through hard times.

Talking about your feelings in a safe place can help you make sense of your past and find comfort. Some people use storytelling to leave a lasting memory or to understand what happened in their life, especially if they are very sick or lost someone they loved.

Others use creative ways, like writing, art, or music, to express their feelings and turn their pain into something meaningful.

In short, storytelling is a powerful way to heal emotionally, mentally, and even physically. It lets you express yourself, connect with others, and find hope in your experiences. It is useful in therapy, self-care, and helping communities heal.

People often share their stories in groups where everyone has been through similar things. This helps create a strong support system for healing. Talking about topics like mental health, illness, or addiction in a safe space helps break the stigma around these issues, making society more accepting.

It is important to note that the effectiveness of therapeutic storytelling may vary from person to person. Some individuals may find it highly beneficial, while others may prefer different therapeutic approaches. The key is to tailor the use of storytelling to the individual's needs, preferences, and therapeutic goals, and it is often most effective when guided by a trained therapist or counsellor with expertise in this area.

Stories about characters going through and overcoming depression or other mental health problems can also help reduce the stigma, give hope, and encourage people to get help. But remember, storytelling is not a replacement for professional help. If you are struggling with depression or mental health issues, it is important to seek help from a mental health expert who can provide personalized care. Therapeutic storytelling can be incorporated into a broader treatment plan to enhance the healing process.

Conclusion

As you can see, therapeutic narratives provide a wellspring of inner strength and resilience, empowering individuals to confront life's obstacles. They are designed to illuminate, educate, instil values, encourage discipline and wisdom, while also supporting problem-solving, inspiring personal transformation, and fostering healing.

Now, as I wrap up this section about the advantages of storytelling, get ready for the real deal – the stories themselves. There are a hundred of them coming your way! Let your imagination, creativity, and morality guide you in deciding when, how, and where these stories leave their mark on you.

PART 2

Cutting the Branch

In a kingdom long ago, there reigned a wise and just king who received a special gift of two magnificent falcons. They were the most beautiful birds he had ever seen.

In his desire to see these magnificent creatures soar gracefully through the skies, the king entrusted them to his most skilled falconer. This falconer was renowned throughout the kingdom for his expertise in training birds of prey, and the king had great expectations.

Months passed, and the king eagerly awaited news of his prized falcons. One day, the head falconer approached the king with a troubled expression. He explained that while one of the falcons had taken to the training splendidly, soaring majestically through the heavens, the other bird remained perched on its branch, motionless and stubborn.

"I've trained countless falcons, Your Majesty," the falconer lamented, "but I've never encountered a bird as obstinate as this one. It refuses to budge, and every attempt to coax it into flight has been met with defiance."

The king, perplexed by this peculiar predicament, gazed at the unmoving falcon. He pondered the matter and decided to grant the falconer more time to work with the stubborn bird, for he believed in the falconer's unparalleled skills.

Months turned into seasons, and still, the falcon clung to its branch, resolute in its refusal to take flight. The king grew increasingly frustrated and sought the counsel of healers and sorcerers from across the realm. These wise individuals attempted various remedies and enchantments to compel the falcon to fly, but their efforts were in vain. The falcon remained rooted to its perch, as though bound by an unseen force.

Desperation consumed the king, and he confided in his trusted minister, saying, "Perhaps we have been approaching this dilemma from the wrong perspective. Maybe we need someone who possesses an innate understanding of nature and animals - a person intimately connected to the countryside."

With renewed commitment, the king issued a royal decree, commanding his court to find a farmer who could provide a solution to this perplexing problem. He believed that someone with an intimate knowledge of the natural world might hold the key to unlocking the falcon's potential.

The following morning brought an astonishing sight - the once-stubborn falcon soared gracefully above the palace gardens, its wings slicing through the air as it joined its companion in the boundless sky. The king, overjoyed by this miraculous transformation, summoned his minister, and demanded to meet the individual responsible for this extraordinary feat.

The minister returned with a simple farmer, a humble man whose clothes were weathered by the toil of the land. The king regarded him with curiosity and asked, "How did you accomplish what skilled falconers, healers, and sorcerers could not? How did you make this falcon take flight?"

With humility and sincerity, the farmer replied, "Your Highness, I am but a simple farmer with limited knowledge, but I understand the ways of the land. I realized that this falcon remained grounded because it had no reason to leave its branch. So, with a simple act, I cut the branch where the bird used to sit. It had no choice but to spread its wings and join its partner in the boundless sky."

The Magpie's Nest

Long ago, only one bird in the entire world possessed the knowledge of nest-building.That bird was the magpie.

One bright day, birds of all kinds from every corner of the land flocked to the magpie and made a humble request, "Dear Magpie, could you please teach us the art of crafting nests for ourselves and our little ones?"

The magpie, ever gracious, replied, "I would be delighted to teach you all, but only if you promise to watch and learn until the end."

The birds eagerly agreed, and the magpie led them to a towering oak tree nestled in the heart of a forest. There, amidst the streak of sunlight filtering through the leaves, the magpie initiated her lesson.

With great care, the magpie began to demonstrate the complex process of nest-building. She started by gathering some mud and shaping it into a round, mud cake.

"Ah, so that's the starting point," remarked the thrush with deep understanding, and off it flew to create nests for thrushes in the same manner.

The magpie continued, undisturbed by the departing thrush. "Now, take some twigs and arrange them around the mud like this," she instructed.

The blackbird observed and exclaimed, "That's how it's done!" before flying away to construct nests for blackbirds.

Still, the magpie persevered, her focus unwavering. "Next, cover the inside of your nest with mud," she directed.

But the wise old owl scoffed, "Who would go to such trouble?" and took flight, leaving the magpie to complete her nest-building tutorial alone. The owl, content with her own unique approach, went on to lay her eggs in a crow's abandoned nest.

Undeterred, the magpie continued to mould the nest. "Now, take more mud and sticks to create a protective dome over the nest," she instructed.

The crow, however, disagreed and remarked, "I have no need for such a dome," remarked the crow as she flew away.

The magpie, now nearing the end of her lesson, lined the outside of the nest with additional twigs, ensuring a sturdy structure. Finally, she adorned the inside of the nest with soft feathers plucked from her very own breast.

So, it went on, every bird taking away a little piece of knowledge and flying away. This is how different birds fashion their nests differently, except for one bird - the turtle dove.

The turtle dove, captivated by the beauty of the completed nest, remained in awe. However, she was, perhaps, a bit too laid-back to undertake the task herself. She cooed contentedly, "Make two, Magpie, make two."

The magpie, feeling a pang of annoyance, retorted, "Why two? One is enough." In a fit of bad humour, she took flight, disappearing into the depths of the forest, leaving the turtle dove to coo peacefully by herself, seemingly content to admire the magnificent nest rather than make her own.

And so, different birds fashioned their nests in their own unique ways, each a testament to the lessons they had learned from the wise magpie.

The Crow and the Peacock

A crow named Tyson led a contented life in a forest. He perched on branches and soared through the skies, relishing the simple pleasures of his avian existence. One day, while Tyson was gliding through the canopy of trees, he caught sight of Cygnet, a graceful swan whose feathers gleamed in pristine whiteness. This vision sparked a thought in Tyson's mind, "Cygnet is so pure and white, while I am so dark and black. Surely, he must be the happiest bird in the world." Tyson decided to share his musings with Cygnet.

Approaching Cygnet with curiosity, Tyson said, "Cygnet, you possess such elegance and whiteness. I can't help but think you are the happiest bird of all."

Cygnet, with a serene smile, replied, "Ah, dear crow, appearances can be deceptive. While I do lead a tranquil life, I too have looked upon other birds and found myself envious. You see, I was once the happiest bird around until I caught a glimpse of a parrot. The parrot, with its vibrant green plumage, is a truly captivating creature. I believe it to be the happiest bird in all of creation."

Intrigued by Cygnet's words, Tyson set off to find Echo, the parrot. Upon reaching Echo's abode, he inquired about the parrot's happiness.

Echo, perched amidst lush foliage, replied, "I have lived a colourful and joyful life, Tyson, until I laid eyes upon a magnificent peacock. My feathers may be green, but the peacock, with its resplendent array of colours, surpasses all. I am convinced that the peacock is the happiest bird of them all."

Driven by curiosity, Tyson journeyed to the zoo where Lapis, the peacock, resided. He observed the throngs of people who gathered daily to admire Lapis' stunning plumage. After the last visitor had departed, Tyson approached Lapis with admiration.

"Dear Lapis," Tyson began, "you are a spectacle of beauty. Thousands come to witness your splendour each day. When people see me, they often drive me away. Surely, you are the happiest bird on this planet."

Lapis, while preening his resplendent feathers, considered Tyson's words, and replied, "I have long believed that my beauty and grace made me the most beautiful and joyful bird on Karth. However, the irony of my situation lies in the fact that my beauty has entrapped me within this cage. As I've observed the zoo carefully, I have come to a profound realization. The crow, Tyson, is the only bird that roams freely, untethered by bars or boundaries. So, in recent days, I have found myself yearning to be a crow, to enjoy the freedom that you possess."

"Cygnet" - a young swan

The Stone Soup

Once upon a time, a traveller who was tired and hungry reached a village. The villages were poor and hid the little supply of food they had from their friends and neighbours. He asked the different people he met about finding a place to eat and sleep for the night. These people kept their doors and windows tight shut and did not like to share.

The traveller stopped at the first house and knocked on the door asking for something to eat. A woman opened the door only a tiny crack. "I am a tired and hungry traveller, Please, may I have something to eat and a place to sleep?"

The woman said, "I'm sorry, I have nothing in the house right now."

The traveller although he was tired and hungry, was not ready to give up. He picked up a large round stone from his satchel and knocked once more at the door. The lady came to the door again, opening it only halfway. "Don't worry," said the stranger. "I have a soup stone. If you will let me, put it in a pot of boiling water, I'll make the most delicious soup in the world. A very large pot, please."

"Yes?" She asked.

"Since you are poor like me, perhaps you too would like to have some of my stone soup."

"Stone soup?" The woman laughed as she looked at the stone in his hand. "You can't make soup from a stone!"

"I've done it before," replied the traveller.

The woman had never seen anyone make soup from a stone before, but since she was hungry too, she invited him in. Then she lit the fire and placed their kettle of water on top and opened the windows to let out some heat. The traveller placed the stone inside the water until it boiled. He sipped a spoon full of hot liquid.

"It's almost done," he said. "But if you had just a little salt and butter the soup would taste so much better."

The woman went to the cupboard and returned with salt and butter. Just as the traveller was pouring them in the pot, the woman's husband returned home. In his hands were carrots and potatoes. What are you making? asked the woman's husband.

"Stone soup." replied the woman and the traveller.

"Impossible!" shouted the husband.

"It's almost finished," the traveller assured the husband as he tasted another spoonful. But it would be even nicer if we added some carrots and potatoes." The husband agreed for he was very hungry and dropped the carrots and potatoes in the pot!

Soon, the smell of the soup drifted out of her windows and down the lane. One neighbour who usually stayed inside wandered out and followed the smell all the way to the first house, where he heard them talking about the stone soup.

"Is the stone soup ready now?" the woman and the husband asked the traveller. "Yes, but it could be even better if we had some turnips and beans! He replied. "I have some," yelled a neighbour who was watching from the window. The neighbour, curious to taste the soup that was made from a stone, returned with the turnips and beans. He dropped them into the pot and the smell drifted even further down the lane.

The news about a strange traveller making soup out of a single stone drew many villagers out of their homes. They followed the delicious smell. "Is the stone soup ready now? asked the villagers when they arrived. "Yes, but I remember having stone soup with chicken and broth in the stew once," he replied.

"I have chicken," said a farmer who ran home to get some.

"I have broth," exclaimed another neighbour who ran to fetch it.

The farmer returned and placed two pieces of chicken into the pot then the other neighbour added the broth. The pot was so full that it almost spilled over. The traveller lifted the spoon to taste it.

“Perfect!” he exclaimed. Then, he served a bowl of stone soup for every single one of the villagers to taste.

“It’s magic!” The villagers cried out seeing how much soup he had made. “Delicious!” cried out another villager. “But where can we get a magic stone? Surely this one has been used up.” The traveller shook his head and pulled the stone out of the pot. The stone was still whole!

The traveller drank the leftover soup and went on his journey. At last, from that day on, the villagers shared what they had with each other.

The Stingy Artist

In ancient Japan, there lived a Buddhist monk named Gessen, known for his extraordinary talent as an artist. However, what set him apart from other monks was his peculiar habit of demanding payment in advance for his paintings, and his fees were exorbitant. This behaviour earned him the nickname - "greedy monk" among the people.

One day, an unusual request came to Gessen. A geisha summoned him to her residence, an unexpected place for a monk to be found. She had a specific request: she wanted a painting done immediately in her presence. Resolute, Gessen agreed but not before asking, "How much will you pay me?"

The geisha, amid entertaining her patron, replied, "I will pay you any sum you ask for, but the painting must be completed right here, right now."

Without hesitation, Gessen commenced his artistic pursuit. He worked diligently, pouring his skill and creativity into the canvas. When the painting was finally completed, he asked for the highest sum he had ever charged, living up to his reputation for greed.

As the geisha handed him his payment, she could not help but voice her thoughts to her patron, "This man is supposed to be a monk, yet all he thinks of is money. His talent is remarkable, but he possesses a filthy, money-loving mind. How can one display the work of such a greedy soul? His art is good enough for my undergarments!"

With that, she flung a petticoat at him, requesting that he paint a picture on it. Gessen, true to his nature, shamelessly accepted the money, completed the painting as per her request, and departed.

Many years later, the truth behind Gessen's seemingly greedy behaviour came to light. It was revealed that he had been secretly preparing for times of crisis. In his home province, a devastating famine would periodically strike, and the wealthy did little to aid the suffering poor. Unnoticed by anyone, Gessen had commissioned the construction of hidden barns filled with grain to provide for the needy during these dire times.

Another reason for Gessen's pursuit of wealth was the deplorable condition of the road that connected his village to the distant city. The road was impassable for ox carts, causing immense hardship, especially for the elderly and infirm who needed to travel to the city. Gessen took it upon himself to have the road repaired, alleviating the suffering of his fellow villagers.

The final reason for Gessen's apparent greed was a profound act of gratitude and devotion. He constructed a meditation temple, fulfilling the unfulfilled dream of his revered teacher. This temple stood as a testament to his love and appreciation for the guidance he had received.

After Gessen had fulfilled these noble endeavours, he abandoned his paint and brushes. He retreated to the tranquil mountains, dedicating himself entirely to a contemplative life. From that day forward, he never painted another canvas again. Gessen's seemingly greedy nature concealed a heart full of compassion, selflessness, and devotion to the welfare of others, leaving a legacy of kindness and generosity.

Jupiter and the Tortoise

In ancient times, whenever the gods had a cause for celebration, they would extend invitations to all the animals, for the gods revelled in the company of creatures from the earth. One day, Jupiter, the god of gods, decided to take a significant step and announced his forthcoming wedding. He sent out invitations to all the animals, inviting them to join in the joyous wedding feast.

The animals received the invitations with delight, for they cherished the chance to attend a divine celebration. Jupiter was known for his generous and hospitable nature, and his parties were legendary. However, amidst the excitement and anticipation, there was one creature who harboured a different sentiment - the tortoise.

The tortoise, feeling rather grumpy and disinterested in such gatherings, decided to forgo the wedding festivities. He confided in his neighbour, saying, "I have little enthusiasm for parties, my friend. They don't bring me joy, and I find no pleasure in dancing and singing."

As the wedding day arrived, it was indeed a grand success, and the celebrations echoed through the heavens. All the invited animals revelled in the festivities, celebrating the union of Jupiter. Yet, amid the merriment, it did not escape Jupiter's divine attention that the tortoise was conspicuously absent.

Determined to understand the reason for the tortoise's absence, Jupiter sought out the slow-moving creature and inquired why he had not graced the celebration with his presence.

With utmost sincerity, the tortoise replied, "Dear god, Jupiter, I hold no love for grand gatherings and festivities. My heart finds solace in the comforts of my own home. For there is no place quite like one's own home."

Jupiter, irritated by the tortoise's response, thundered with displeasure, "If you hold your home in such high regard, then so shall it be. From this moment forth, I decree that you, tortoise, shall carry your home upon your back for all eternity!" And so, it came to pass that, since that day, tortoises have borne the weight of their homes wherever they go, a constant reminder of their preference for the comforts of their own dwellings over the revelry of grand celebrations.

The Hungry Mouse

In a time of hardship and scarcity, a mouse found himself very hopeless. Food was scarce, and he scoured the surroundings in search of sustenance, but every nook and cranny turned up empty. As days passed, the mouse grew thinner and weaker, a shadow of his former self.

One day, his luck took an unexpected turn. As he continued his desperate quest for food, he stumbled upon a basket brimming with golden kernels of corn. The sight of the corn filled his heart with excitement, and he couldn't wait to satisfy his hunger. Determined to access this treasure trove of food, the mouse scoured the basket's surface for an entry point. Eventually, he discovered a small hole in its side, just big enough for him to squeeze through.

With cautious excitement, the mouse made his way into the basket, and his tiny heart leaped with joy as he began to feast on the corn. Bite by bite, he filled his belly until it was comfortably full. The mouse was content, and the thought of returning home crossed his mind.

However, as he attempted to exit through the same small hole, he realized he was stuck. He felt a sudden rush of panic. He called out for help, tears welling up in his eyes.

It wasn't long before a wise rabbit happened upon the distressed mouse. The curious rabbit asked the mouse what was causing its distress. The mouse explained how he had become trapped inside the corn-filled basket.

The rabbit couldn't hold back a chuckle and remarked, "Your situation stems from gluttony. If you wish to escape, you must wait until you've slimmed down to your previous size."

Accepting the rabbit's advice, the mouse decided to wait. He settled on top of the corn and drifted off into slumber. Time flowed slowly as he slept, and when he eventually awoke, he realized that his belly had indeed shrunk back to its normal size. Overjoyed and relieved, the mouse decided it was time to leave and finally return home.

However, temptation beckoned once more. Greed overcame him, and he resolved to have just a little more corn before departing. He feasted once again until his belly was full to bursting. Yet, when he attempted to exit the basket a second time, he was met with the same stubborn obstacle - the small hole.

The mouse's panic surged, and he came to the disheartening realization that he would have to wait even longer to escape. This time, fate had different plans. A plump and cunning cat was passing by the basket. When the cat saw that there was a mouse inside, she quickly peered in. Without hesitation, the cat seized the mouse, ending his ordeal and sealing the fate of the hungry but greedy mouse.

Deer Hunting

Once upon a time there lived a hunter renowned for his extraordinary ability to mimic the voices of the creatures that roamed the woods. He had mastered the art of deception, using this skill to stalk and capture various animals for his sustenance. His method was simple yet incredibly effective: he would imitate the calls of the forest inhabitants, luring them into a false sense of security before swiftly dispatching them with his arrows. This clever trick never failed him.

One fateful day, the hunter ventured into the heart of the forest with the intention of hunting deer. He carefully selected a concealed spot and commenced his performance, mimicking the graceful calls of the deer. It wasn't long before his imitation echoed through the woods.

However, as the alluring sound of the deer's call travelled through the forest, it caught the attention of an unexpected visitor - a cunning and opportunistic wolf. The wolf, lured by the promise of an easy meal, promptly arrived on the scene. The hunter's heart raced as he saw the wolf approaching, realizing that he was in a perilous situation.

Thinking quickly, the hunter decided to change his tune. With a gulp of nervousness, he imitated the fearsome roar of a tiger, hoping to scare the wolf away. The wolf, hearing this terrifying sound, was overcome with fear, and promptly retreated, vanishing into the undergrowth.

The hunter let out a sigh of relief, but his respite was short-lived. Just as he began to relax, he heard the unmistakable sound of a real tiger approaching. Panic surged through him as he realized he had inadvertently attracted a genuine predator. Desperate to survive, he decided to mimic the low, menacing grunt of a bear. He knew that even tigers were wary of bears and would flee their presence.

Miraculously, the tiger, upon hearing the bear's grunt, turned, and swiftly disappeared into the forest, leaving the hunter trembling but alive. However, his relief was once again fleeting, for now, he had exhausted his stock of animal calls.

Moments later, as if drawn by some malevolent force, a massive bear emerged from the shadows. The hunter's heart sank as he realized that he had run out of tricks. He was now completely defenceless. The bear, driven by hunger and instinct, pounced on the hunter, tearing him to pieces and feasting on his remains.

In the end, the hunter's own cunning had led to his downfall. His mastery of animal mimicry, which had served him so well in the past, ultimately became the instrument of his death in the unforgiving depths of the forest.

The Bee Sting

In a distant realm, there ruled an emperor, Hikaru Yuki, who presided over a vast empire, supported by a mighty army. The emperor was blessed with seven children, all of whom, except for the eldest, Taro*, were known for their obedience and good manners. Taro, on the other hand, possessed a fiery temper that often flared without warning, causing concern for the emperor as he was to be the heir to the imperial throne.

To mould Taro into a wise and composed future leader, the emperor entrusted the education and guidance of his children to Kenzo, a renowned scholar celebrated for his vast knowledge and wisdom. Despite Kenzo's repeated admonitions and teachings, Taro persisted in his unruly and disagreeable behaviour.

One day, during one of Taro's tempestuous tantrums, Kenzo decided to take a drastic step. He firmly grasped the prince and led him to a lush flowering bush, where he forcefully pressed the boy's hand against a cluster of busy bees until one of the bees delivered a painful sting.

The searing pain of the bee's sting was so unexpected and shocking that it brought an immediate end to Taro's furious outburst. Clutching his throbbing hand, he angrily exclaimed to his teacher, "I will inform my father about this brutality!"

Kenzo responded calmly, "When you speak to your father, share this lesson with him."

Curious and puzzled, Taro inquired, "What lesson?"

Kenzo directed Taro's attention to the bee they had just observed. Together, they gazed at the agonized insect writhing on a leaf, its entrails torn out by its own stinger, until it ultimately succumbed to its pain and perished.

With a solemn tone, Kenzo explained, “That, young Taro, is the cost of anger.”

That night, Taro recounted the entire experience to his father, the emperor, Hikaru Yuki. The emperor, recognizing the profound wisdom in the lesson, rewarded Kenzo with a generous gold piece.

As Taro matured into adulthood and eventually ascended to the throne, he became renowned for his composed judgement and his steadfast refusal to be provoked by anger or agitation. This invaluable trait served him well throughout his long and challenging reign, as he guided his empire through turbulent and uncertain times with wisdom and grace.

"Hikaru Yuki" - "Hikaru" means shining or radiant and "Yuki" means courage. This name implies a bright and courageous ruler.

"Taro" - endearing way to refer to eldest son in Japanese culture.

"Kenzo" signifies wise or intelligent.

Waiting for the Angel

In a vast floodplain, there resided a holy monk who had made this place his humble abode. This serene landscape, however, could transform into a perilous zone during the monsoon season, as the waters swelled, and the surroundings became treacherous. Despite the repeated warnings from concerned neighbours and the rising floodwaters, the monk remained steadfast in his decision to stay within the confines of his simple hut.

To the many would-be rescuers who implored him to evacuate, he calmly replied, "I have received a divine promise in a dream. An angel assured me that he would protect me."

With unwavering faith in the angel's words, the monk watched as his neighbours left their homes in search of safer ground. He observed the local officials arrive in carts, extending offers of assistance and evacuation. Even a dedicated rescue team arrived in a sturdy boat, all with the same message: "You must leave, for the floodwaters are rising dangerously."

Yet, the monk, with serene confidence, turned them all away, trusting in the angel's promise.

However, as the monsoon's relentless flood continued, the inevitable transpired, and the holy monk was tragically drowned by the surging waters. In the moments following his passing, the first being he encountered was none other than the angel from his dreams.

With stern authority, the angel demanded, "What are you doing here?"

Confused and taken aback, the monk stammered, "But you promised me...."

The angel interrupted, exasperated, "I kept my promise, you thick-headed old fool! I sent your neighbours, a cart, and a rescue team in a boat."

The monk finally grasped the truth and simply replied, "Oh!"

In the end, the monk learned a valuable lesson about the nature of divine assistance. Sometimes, help arrives through the actions of ordinary people and the resources they provide. The monk's unwavering faith in his dream was admirable, but he failed to recognize that the assistance he sought had already been offered through the compassionate efforts of those around him. The story serves as a reminder that divine intervention often works through the hands and hearts of those who care and act to help their fellow beings in times of need.

The Broken Dishes

Yuto was a rich landlord who lived long ago in a far-off country. He had a lot of possessions. He had a collection of precious and rare items in his house. Among them he had a set of ten porcelain dishes. He liked it and praised it very much.

Yuto's maidservant Sakura took care of all her master's rare collections. Sakura loved her work and was always loyal to her master. But one unfortunate night Sakura broke one of the much-praised porcelain dishes. Her master punished her severely much to her shock and bewilderment. The heartbroken Sakura threw herself in the well and drowned.

Everyone in the household was awakened one night after Sakura's funeral by the ghastly sound of counting one through nine. Yuto noticed the counting resounded from the well where Sakura drowned. The ghostly sound was followed by silence. When everyone went back to sleep, the counting would begin again.

This continued for several nights. Servants left the house scared. Yuto sought the help of an exorcist priest. The exorcist priest explained that the ghost of Sakura haunted the master to remind him that despite the loss of the single precious piece, the master still retained nine others.

"What must I do to stop this infernal haunting?" begged Yuto. "No one can sleep, and my servants are running away."

"Break the remaining nine pieces and throw them into the well," replied the priest.

"All nine?" inquired the master. "Couldn't I keep just one?"

"No," replied the exorcist. "You have nine and now you will have none if you want to sleep."

Yuto hesitantly did as the exorcist priest commanded, and the haunting stopped. The master was never able to replace the precious set and the servants who fled never returned.

"Yuto" means rich, calm or courageous

"Sakura" means cherry blossom in Japanese.

The Bear and the Fox

In a dense mountain forest, there lived a clever and resourceful fox named Rena. Her days were filled with the pursuit of food and treasures to make her den a snug and inviting home.

One sunny morning, a massive brown bear named Jasper happened upon Rena's den and decided it was the perfect spot for a nap. He settled down, unknowingly blocking the entrance with his massive form. When Rena returned, she attempted to squeeze her way past the bear, but his size proved insurmountable. She decided to wait, thinking that Jasper would surely move on the next day.

However, day after day, Jasper returned to the same spot, leaving only when Rena was sound asleep. The fox's attempts to politely find a way around the bear proved fruitless.

As time passed, Rena's den began to deteriorate, and her family grew hungrier with each passing day. The once-comfortable home had now become a symbol of their predicament.

One day, as Rena's cubs nestled close to her, one of them spoke up, "Mama, why don't you just ask the bear to move?"

Rena pondered this idea. The next morning, when Jasper arrived as expected, she gathered her courage, cleared her throat nervously, and popped her head out of the den. "Um, excuse me," she began, "would you mind moving? You see, you're blocking the entrance to our den."

Jasper, surprised by Rena's polite request, looked down at the fox with raised eyebrows. "Oh my, I had no idea," he replied with genuine remorse. "You should have told me days ago. I'm terribly sorry."

With that, Jasper promptly relocated to a nearby spot for his daily nap, freeing up the entrance to Rena's den. Rena and her family wasted no time. They worked together to hunt for food, gather supplies, and repair their den. Before long, they were enjoying hearty meals and living in the coziest den imaginable.

All this transformation had come to pass because Rena had finally summoned the courage to take matters into her own paws and politely ask Jasper to move. It was a lesson for all the creatures of the forest that sometimes, a simple conversation could resolve even the most daunting of problems.

Glimmers of Insight

In an ashram, there lived a devoted seeker who was the cherished disciple of a great Guru. Day after day, he sat at the revered feet of his teacher, absorbing the profound wisdom that flowed from the Guru's lips.

The ashram was a place of pilgrimage, attracting numerous visitors from far and wide, all seeking the guidance and teachings of the enlightened Guru. It was a customary practice that the Guru often engaged his visitors and disciples by sharing profound lessons through the art of storytelling.

One day, one of the eager disciples, burning with curiosity, ventured to ask a question that had long lingered in the hearts of many: "Guruji, why do you choose to convey your teachings through stories? Why not simply impart your wisdom directly to us? Would it not be a quicker and more direct path to enlightenment?"

The Guru, sitting in serene contemplation, looked upon his disciple with a knowing smile. He gestured for the disciple to approach, and in a calm and gentle voice, he replied, "Bring me some water."

The disciples were familiar with their Guru's disciplined and formal conduct. He had never asked for water at this hour of the day before. Nevertheless, they respected his words and promptly obeyed his request. One of the disciples, a sincere and obedient soul, hurried to the kitchen and fetched a clean brass water pot. He walked to the well, drew water with care, and returned with the filled pot, offering it to his revered teacher.

With a hint of amusement in his eyes, the Guru, holding the pot, asked his disciple, "Why have you brought me a pot when I asked only for water?"

The disciple, taken aback by the question, was momentarily puzzled. He stammered, “Guruji, I... I thought you wanted water, so I brought the pot to carry it.”

The Guru nodded and gently explained, “Just as you brought a pot to carry water, I use stories as vessels to carry the wisdom of life. The stories are the vessels, and the teachings are the water they hold. They make the teachings easier to grasp and remember, allowing them to flow into the depths of your hearts. Just as you would find it challenging to carry water without a pot, the stories provide a vessel for the profound truths of life to be carried and cherished.”

In that moment of clarity, the disciple understood the profound significance of the Guru’s storytelling. The stories were not merely tales but vessels that transported the precious wisdom of life itself, making it accessible and memorable for all who sought the path of enlightenment.

The Frozen Bird

Once upon a time there lived a young sparrow, Robin. He was very active, possessed boundless energy but was equally stubborn. When winter descended upon the land and his fellow birds began their journey southward, Robin resisted the call of migration. The older, wiser birds warned him of the perils of staying behind: he would not withstand the trials of a northern winter, the dangers of facing predators alone or freezing to death. Yet Robin remained resolute, bidding farewell to the last of the departing flock. He wanted to stay right where he was.

The initial days were challenging, and Robin struggled to fend for himself. He believed he could manage alone, but as the temperature continued to drop, he understood the gravity of his decision. Winter's icy grip grew relentless, and Robin finally realised that it was time to follow his fellow birds to the south.

However, as he ascended into the chilly sky, snowflakes began to descend upon him. They weighed down his fragile wings, rendering him powerless to continue his flight. Robin's body grew numb, and he fell to the ground below.

As he lay on the frozen ground, despair setting in, Robin began to pray for a miracle. His hope was fading fast. Just when he believed that survival was impossible, a cow happened upon the scene and dropped a pile of dung on him.

The dung, though initially shocking and unpleasant, had an unexpected effect. The little sparrow could hardly breathe and thought for sure he would be buried alive. Within minutes, he was unfrozen, able to breath easily. He was so happy to be alive, warm, and breathing easily that he started to sing for joy.

His cheerful melody caught the ear of a passing cat, arousing its curiosity. The cat followed the sound and soon discovered the source - a little sparrow buried beneath a pile of cow dung.

Without hesitation, the cat unearthed the helpless bird and saw in Robin a delicious meal. The unfortunate sparrow's miraculous escape from the cold was followed by an equally unexpected death at the jaws of the cat. Thus, in a twist of fate, Robin's survival turned into his ultimate downfall.

Yuri's Redemption

In a calm Ukrainian village, there lived a humble farmer and his loving wife, along with their faithful sheepdog, Yuri. The devoted dog had spent countless years serving the farmer and his family, herding their sheep with unwavering loyalty. However, the toll of time had rendered Yuri older and slower, and he could no longer perform his duties with the same vigour as before.

One chilly evening, as Yuri lay by the warm crackling fire, he overheard a hushed conversation between the farmer's wife and her husband. She whispered, "Yuri is growing too old for his work. It's time we found a way to part with him."

The farmer, who cherished Yuri deeply, tried to persuade his wife to keep their ageing friend. He argued, "Yuri has been a faithful companion all these years. He deserves a peaceful retirement."

However, the farmer's wife was resolute. She explained, "As much as I appreciate Yuri's loyalty, we need a younger, more agile dog to tend to the sheep, especially now that we have a newborn baby. I can't help you chase after the sheep if they stray. Yuri must go."

Yuri felt a heavy sadness wash over him as he realized the inevitable fate that awaited him. There seemed to be no place for him anymore, no home to return to. The entire night, he wrestled with his emotions, summoning the courage to face an uncertain future.

The following morning, the farmer led Yuri deep into the forest. His love for the dog was so profound that he couldn't bear to take Yuri's life. Instead, he left him there, hoping that Yuri might find a new life in the wilderness.

As Yuri roamed the unfamiliar forest, a curious wolf crossed his path. The wolf was intrigued by the sight of a domesticated dog in the wild and inquired, "What brings you here, wandering alone?"

Yuri, his voice filled with melancholy, replied, "I have been cast aside by my master. He believes I am too old to serve them."

The compassionate wolf offered a lifeline, saying, "I can assist you if you're willing to listen. If you follow my guidance, your master may take you back."

Yuri's heart leaped with hope, and he implored, "Please, my dear friend, help me."

"Don't you worry" said the wolf, "Your master, you know, goes out every morning very early with his wife into the field; and they take their little child with them, and lay it down behind the hedge in the shade while they are at work. I will come out of the wood and run away with the baby. You must run after me as fast as you can, and I will let it drop. Then you may carry it back, and they will think you have saved their child, and will be so thankful to you that they will take care of you as long as you live."

Yuri liked the plan, and it was so well managed. The wolf, as promised, made off with the baby, and the farmer and his wife gave chase, terrified for their child's safety. Yuri pursued the wolf, and before long, the cunning dog defeated his wild counterpart and safely carried the baby back to its frantic parents.

Tears of relief filled the farmer's eyes as he patted Yuri's head and declared, "Old Yuri has rescued our precious child from the wolf ! He shall live with us for the rest of his days and want for nothing."

From that day forward, Yuri lived a life of comfort, cherished by his master and mistress, grateful for his second chance at a life filled with love and care.

The Baker's Daughter

Long ago, in a small town in England lived a baker who was greedy. He sold the best bread in the town. But he was mean and stingy. He cheated people in every way possible. Every night he would sit at his table and count how much money he had made that day, repeatedly. He would decide to make more money the next day.

The baker had a beautiful daughter, Lucy, who helped him in the shop, and she learned all the tricks of the trade and became more cunning and mean than her father.

One day, the baker went to another town leaving his daughter in charge of the shop. When she was working alone, an old woman came into the shop. She was in rags, looking tired and famished.

The old lady said, "Would you be so kind to give me a bit of your dough? I haven't eaten for three days."

Lucy was reluctant to give anyone anything for free. But the old lady wouldn't go away, and the girl thought that the presence of the woman in rags and arguing with her would be bad for business. Finally, she agreed and gave the woman a tiny bit of dough, just to get rid of her.

The old woman took the piece of dough, but then she asked Lucy, "Please, my dear girl, would you be so kind as to bake this tiny bit of dough in your oven? You see, I don't have an oven, and I'm so terribly hungry."

"Fine," the girl said, "give it to me." She took the tiny bit of dough and put it in the oven.

The old woman sat down and waited.

After a while, Lucy went to check on the dough, expecting to see the world's tiniest loaf of bread possible. But the tiny piece of dough had turned into a huge loaf of bread!

The greedy girl couldn't believe her luck. Quickly, she took the bread out of the oven and hid it in a basket. Then she turned to the old woman.

"I couldn't find your bread in the oven. It must have fallen through into the flames!"

The old woman said, "Could you please try once more, dear? I can wait a little longer."

Lucy didn't know what else to do, so she took another little bit of dough, even tinier than before, and put it in the oven while the old woman waited. But when the girl looked in the oven again, she discovered another loaf that was even larger and more appetizing than the first.

"Oh, you wouldn't believe this," Lucy called out to the old woman. She pretended to look about the oven as if in search of something. "I cannot find the cake," said Lucy. "It must have tumbled into the fire and got burnt again. It's gone!" And she hid the second loaf also in the basket.

"Well, my dear, maybe the third time will be lucky. Can we try one last time? I beg you. I'm starving!" said the poor old woman.

This time, Lucy used a piece of dough so tiny, it was almost invisible. She put it in the hot oven, and once again it turned into an inviting, huge loaf of bread.

This time, however, the old woman had stood up so she could have a good view of the kitchen, and when she saw the girl trying to hide the loaf in the basket, she shouted: "Hey! That's my bread!"

Lucy protested, but of course the old woman knew about all three of the loaves the selfish girl had stolen for herself.

Lucy did not know that this old woman was a fairy in disguise! The fairy decided to punish the girl for her greed and meanness. She took her magic wand out of her cloak and quickly tapped the girl's back three times.

And lo! Lucy turned into an owl! The bird flapped around the bakery for a moment, then flew outside through the open door, and no one since has ever seen the baker's daughter again unless you are up late at night and hear a hoot. It's the baker's daughter, saying, "I'm sorry."

And the Rooster Crowed

Once upon a time, in the town of Barcelos, in Portugal, a wandering pilgrim by the name of Hugo was on the way to Santiago de Compostela in neighbouring Spain. However, an unsettling cloud of suspicion hung over the town, for a grave crime had been committed, and the culprit was not found. Unfortunate for Hugo, he happened to be the sole stranger passing through Barcelos at the time, casting him as the prime suspect in the eyes of the cautious townsfolk.

Despite his vehement protests and repeated declarations of innocence, Hugo was swiftly charged with the crime. To the residents of Barcelos, he was no longer a pilgrim on a spiritual journey but a convenient scapegoat. The wheels of justice turned mercilessly, and Hugo found himself condemned to death by hanging.

As the day of his impending execution drew near, Hugo harboured one final wish - to speak with the judge responsible for his grim fate. His request was granted, and he was led to the judge's grand residence, where a lavish banquet was underway, attended by the judge's family and closest friends.

Hugo made the last attempt to prove his innocence. Everyone at the banquet ridiculed him. He was heartbroken and in a moment of despair Hugo pointed to the table where there was a roasted chicken and in a trembling voice, he declared, "As certain as I am innocent, so too is it certain that this rooster will crow when they hang me!"

The assembled guests, however, met his proclamation with mockery and scorn. Hugo's heart sank as he faced their ridicule, feeling the weight of despair pressing upon him.

The day of reckoning arrived, and Hugo was led to the gallows in the town square. The impossible happened before the incredulous onlookers' eyes - the rooster, perched on a nearby fence, suddenly stood tall and gave out a mighty crow, breaking the silence of the moment.

There could no longer be any doubt about the pilgrim's innocence. The judge, astounded by this extraordinary turn of events, rushed to the gallows. To his astonishment, Hugo was indeed still alive, the noose around his neck miraculously slackened, preventing him from being strangled.

Released from the clutches of death, Hugo wasted no time and resumed his journey toward Santiago. A few years later, filled with gratitude for this unexplainable miracle, he returned to Barcelos. There, he sculpted a lasting monument - a magnificent stone cross, dedicated to the Lord of the Rooster, a symbol of the incredible turn of fate that had spared his life and cleared his name.

The Happy Family

Once upon a time, in a dense forest, there lived a charming and tightly knit family of squirrels. This family, known as the Eichhorns, consisted of Mr. and Mrs. Eichhorn*and their three adorable children. They resided high up in an oak tree, at the very heart of the forest. Their food was simple, their nest poor, yet they were happy and loving. They spent their days scampering through the trees, leaping from branch to branch, and performing daring acrobatics that left other forest creatures in awe.

One day, Mr. Eichhorn told his wife, "I feel so sad to see you and our kids living in such poor conditions. I want to provide you the ripest, juiciest fruits and nuts and a very comfortable nest."

Mr. Eichhorn worked hard day in and day out. He came home late and hardly had the time to speak to his wife and children. He gathered more fruits, more grains, and more nuts. He wanted to offer his wife and children the very best in the world. He became the richest squirrel in the woods. But he was never satisfied. He wanted to have more and more.

Their nest was the biggest and the brightest for miles around. But the three baby squirrels missed their dad terribly and felt very lonely. They could never see their father, no one could hear his voice, nor feel the warmth of his touch. He was forever out to gather more wealth.

One day the little ones asked their mother, "Where is Daddy? He is never with us! We feel so lonely. Does he not care for us?"

With tears in her eyes, the mother squirrel replied, "No. Daddy loves you very much. He works very hard to get you better food, more money, and more toys. He wants to give you the best." She knew how much they missed their father's playful antics, his daring adventures, and his bedtime stories under the twinkling stars.

"The little ones said, "Mom, we want Daddy. We want to see him, hear him, play with him. Without him we feel lonely, we miss him."

That night when father squirrel came home in the small hours of the morning, she could not help telling him, "Darling, I feel lonely, the children feel lonely. We don't feel loved and cared for. We need you. We want your love; the children want your love. We do not want more money. But we want you."

Mr. Eichhorn was baffled. In a temper he shouted, "What? What are you saying? That I don't love you? I am working myself to death for you! I want to give you the best house, the best food, the best clothing, the best of everything, and now you have the cheek to tell me that I do not love you! For whom am I doing all that? How ungrateful you are!"

Mrs. Eichhorn replied, "It is not a palace our children need but a happy home! It is not money, food, comfort we long for. I long to be with my husband, and the children are craving for a father. Can't you see that?"

Mr. Eichhorn yelled, "You women don't seem to understand. Always asking for love and affection! I wanted to give you a bright future, economic security, a life free from worries."

This argument between the husband and the wife awakened the children. Only today, the youngest one could see his father's weary face for the first time.

The eldest whispered to him, "You see, this man is our father. He works very hard, he's the one who buys us everything we have."

The youngest one innocently whispered back to him, "Then, tell him to buy us a Daddy."

"Eichhorn" is a German surname that means "squirrel" in English. It is a relatively common surname in German speaking countries and used historically to describe someone who has characteristics associated with squirrel, such as nimble or agile.

Feeding on the Future

In a grove near a village, a flock of birds led a leisurely life. Their days were filled with carefree flutters, searching for food, and sipping from clear streams, as sustenance was abundantly available in the nearby village. The villagers were generous, and the birds had never known hardships.

In the village, there lived a rich farmer, Samuel, who owned a large field. He worked hard in his field and did all the work there together with his family and friends. He ploughed the soil, nourished it with rich manure, and irrigated the fields meticulously. Then he bought the best seeds and scattered them across his well-prepared land with a generous hand. He toiled very hard all day long and, in the evening, when he saw his field, he was content and hopeful for a bumper harvest.

From their lofty perches in the trees, the birds observed Samuel's tireless efforts. They were baffled by the farmer's act and exclaimed, "What a fool! He is throwing away all those wonderful grains. Doesn't he know that the best grains are for eating? What a waste!"

The next day, all the birds gathered and had a conference on how best to salvage the seed that the wasteful farmer scattered all over the fields. "Let us go to the field and enjoy those seeds," they said. "Not a grain should be lost."

In an instant, the field was swarmed by birds of all kinds. They pecked and scratched at the earth with fervour. Those with strong claws made quick work of the task. Birdsong from a thousand throats rent the air. There was chirping and singing all over the fields.

Time passed, and the seasons turned. The rains came and went, the sun completed its daily journey countless times, and the land experienced the ebb and flow of seasons. Yet, the once-fertile field remained barren. Only a handful of timid cornstalks dared to raise their heads here and there. The crop had failed, and Samuel was left perplexed, despite his hard labour and the finest seeds sown.

The hungry birds, now emaciated and disheartened, exchanged concerned whispers. "What went wrong this year?" they lamented. "We face the prospect of starvation!" They did not know their plight was the consequence of the grand feast they had enjoyed during the summer.

In their haste to seize the seeds, they had unwittingly disrupted the carefully planned cultivation process. Samuel's labour had been in vain, and the once-thriving field had been left barren by the very creatures that had sought to benefit from it. In their pursuit of instant gratification, the birds had unwittingly eaten up their future harvests.

Friend of the Same Size

Once there lived a friendly and amiable elephant, Gauri. She lived happy and contented in the dense and vibrant jungle among her fellow elephants within her close-knit herd.

One splendid, sunlit day, Gauri ventured away from her herd, lured by curiosity and a desire for exploration, and soon found herself wandering deep into an unfamiliar forest. As she treaded farther into this strange and enchanting territory, she realized that she had become lost, leaving her bewildered and yearning for the companionship of her beloved friends.

Gauri was a social creature who thrived on the laughter and giggles shared with her dear companions. However, in this unfamiliar and unknown jungle, she felt an overwhelming sense of strangeness and uncertainty. She gazed around, her eyes searching desperately for the familiar faces of her friends, but her efforts proved fruitless. Worry, fear, and loneliness gripped her heart.

Desperate to alleviate her solitude, Gauri made loud trumpet calls in the hope of summoning her friends, but alas, no familiar faces appeared. Realizing that her cries went unanswered, she decided to embark on a quest to make new friends in this unfamiliar land, as she roamed throughout the jungle in search of potential companions.

As Gauri ventured through the undergrowth, she spotted a mischievous monkey perched on a tree, happily munching on ripe bananas. She approached the monkey and initiated a conversation, "Hello there! I'm Gauri, and I am new to this jungle. May I inquire about your name?"

The monkey, who went by the name of Kapi, responded, "My name is Kapi. What brings you here?"

With a longing for companionship, Gauri continued, "I'm feeling quite lonely in this unfamiliar place. I yearn for friends to share moments of play and conversation. I've been watching you, and you seem like a fun and lively companion. Would you consider being my friend?"

Kapi, the monkey, didn't hesitate to reply, "You are much too large to swing from tree to tree as I do. How could we have fun together? I'm sorry, but I can't be your friend."

Disheartened by the rejection, Gauri moved on, continuing her search for new friends. Her quest led her to a delightful sight - a rabbit joyfully hopping in and out of the bushes and bouncing around. Gauri watched from a distance, captivated by the playful display. Drawing closer to the rabbit, she greeted it with a warm smile.

"Hello there!" Gauri exclaimed, her spirits lifted, "My name is Gauri, and I am a newcomer to this jungle. May I know your name?"

The rabbit, looking up at the colossal elephant before it, replied, "Hello! I am Bunny, and I am a rabbit."

Curious about Gauri's intentions, Bunny inquired, "What can I do for you?"

Gauri replied with enthusiasm, "I am lost in this unfamiliar land, and the pain of loneliness weigh heavily on me. I am in search of new friends. I've been observing your lively antics, and you appear to be a delightful companion. Would you consider being my friend?"

However, Bunny took a thoughtful look at the giant elephant and expressed, "You cannot fit into my burrow; you're too large. I'm afraid I can't be your friend."

Gauri's disappointment grew, but she continued her journey, eager to find someone who would be willing to be her friend. Soon, she encountered a frog named Green, leaping between the land and water, exhibiting incredible agility.

"Hey there, frog! I'm Gauri, and I am new to this jungle. Would you be my friend?" Gauri inquired eagerly.

The frog, named Green, looked up and acknowledged her presence. “You are too large and heavy to leap like me,” Green responded. “I’m sorry, but I can’t be your friend.”

Gauri’s quest to make friends led her to meet Billy the bear, Shikari the fox, and various other creatures along her path. Regrettably, she encountered the same response from each one. No one seemed willing to be her friend due to her considerable size.

With each rejection, Gauri’s heart grew heavier, and she continued to feel desolately alone in this strange forest. The days passed, and her search for companions ended in disappointment. She longed for her old friends who had cherished her presence.

One day, Gauri observed a sudden panic among all the animals. They were fleeing deeper into the forest, their eyes filled with fear and terror. Gauri wondered what had instilled such dread in her fellow creatures. She spotted Billy the bear running in the distance and hurried to inquire about the commotion.

Billy, panting and clearly alarmed, shared the reason for their panic. “The lion is on the loose, and we are all fleeing for our lives!”

The lion, it appeared, was on a merciless rampage, gobbling any creature that crossed his path. Gauri was stunned and deeply concerned for the well-being of the animals in her newfound home. She made a courageous decision to confront the lion and attempt to reason with him. Her immense size made her an unlikely target for the lion’s aggression.

With unwavering determination, Gauri approached the lion and spoke firmly, “Please, sir, spare these innocent animals. Let them be.”

The lion, his eyes filled with anger, regarded the colossal elephant before him but chose to disregard her presence.

Gauri persisted, “They are my friends, and they are entirely blameless. I implore you not to harm them.”

In response, the lion scoffed at her and advised her to step aside and refrain from interfering in his affairs. The lion's arrogance and indifference only served to further ignite Gauri's indignation.

Infuriated by the lion's callousness, Gauri, in a burst of rage, delivered a powerful kick, sending the lion soaring through the air until he collided with a rock, injuring himself severely.

Having subdued the lion and returned to the other animals, they emerged cautiously from their hiding places and began to celebrate their newfound safety. Gratitude filled the air, and they collectively voiced their appreciation to Gauri, remarking, "You are just the right size to be our friend!"

Gauri had, at last, secured the friends she had longed for, friends of all shapes and sizes who cherished her for her courageous and caring spirit.

"Kapi" is a word for monkey in Sanskrit.

The Donkey's Brain

Once upon a time in a jungle, there lived a lion named King Leo. He was the undisputed ruler of the animal kingdom. As the years passed, Leo's age began to catch up with him. His once-mighty legs grew feeble, and hunting for food became increasingly difficult. Starvation loomed, and Leo knew he needed someone to hunt for him.

Leo believed that a fox named Max was the ideal candidate for the position. He admired Max's intelligence and wit and summoned him, saying, "Dear friend, I've always valued your cleverness. I wish to appoint you as my assistant and counsellor for all matters in the jungle."

Max, though sceptical of Leo's intentions, couldn't refuse the king's request. He pretended to be delighted and replied, "Your Majesty, I am deeply honoured by your trust and gladly accept your offer."

Leo was pleased with Max's response and quickly dispatched him to find an animal for his meal. On his journey, Max encountered a plump donkey named Sharp, grazing peacefully in the forest. Max approached Sharp and exclaimed, "Friend, where have you been hiding? I've been searching for you tirelessly."

Curious, Sharp asked, "Why were you looking for me? What's the matter?"

Max smiled and said, "You are a very lucky donkey, my friend. An incredible honour awaits you. King Leo has chosen you to be his chief minister, and he asked me to deliver this wonderful news."

Sharp, bewildered by the unexpected offer, stammered, "King Leo? I'm afraid of the lion. He might eat me. Why would he choose me as his chief minister? I'm not fit for such a position. Please, leave me be."

Max chuckled and reassured Sharp, "You underestimate yourself, Sharp. King Leo has heard of your remarkable qualities. He's eager to meet you because you are wise, gentle, and hardworking."

With Max's convincing words, Sharp began to consider the possibility that he had underestimated his own potential. He cautiously agreed to accompany Max to meet the lion.

Leo welcomed them warmly, but Sharp couldn't hide his nervousness. As they approached, Sharp hesitated, refusing to come closer to the hungry lion. Max, quick on his feet, informed Leo, "Your Highness, our chief minister appears to be a bit shy and hesitant."

Famished and impatient, Leo lunged at Sharp but couldn't overpower the sturdy donkey. Sharp escaped, and Leo was left empty-handed.

Infuriated, Leo roared at Max, "You've deceived me! I am starving. Bring that donkey back, or I'll devour you."

Max replied, "Forgive me, Your Highness. You were too hasty. You should have left it to me to approach him carefully. I'll make another attempt."

Max found Sharp once more and persuaded him to return to Leo, emphasizing the power and prestige that awaited him as the lion's second-in-command.

When they faced Leo again, the lion was more careful. He greeted Sharp with a smile and said, "Welcome, my good friend. It was unkind of you to run away. Come closer; you are now my chief minister."

With a sense of trust, Sharp approached the lion, who seized the opportunity and swiftly struck him on the head, ending his life. Leo thanked Max for his cunning and, at last, had a meal to satisfy his hunger.

As Leo prepared to feast on the donkey, Max said, "Your Majesty, it is customary for a king to bathe before a meal." Leo, recognizing the wisdom in this advice, left to take a bath, leaving Max to guard the donkey's carcass.

While Leo was away, hunger gnawed at Max, and he couldn't resist. He decided to enjoy a portion of the donkey's brain, considering himself the rightful recipient.

When Leo returned and saw the donkey's head had been opened, he asked, "What happened to the donkey's head?"

Max, feigning innocence, replied, "Your Majesty, you were so eager to eat that you smashed the donkey's head when you killed him."

Leo was satisfied with the answer and began to eat. "What happened to the donkey's brain? I wanted to eat the brain first", he roared in great anger.

"The donkey's brain!" Max expressed his surprise. "Your Majesty, you're fully aware that donkeys don't have a brain. Had that donkey ever had a brain, he would never have come with me here a second time."

"Yes," agreed Leo, "You are right."

King Leo, satisfied with the response, continued his meal, unaware of the trickery that had transpired.

Feasted to Farewell

Once upon a time, in a far-off kingdom, there lived a rare and magnificent sea bird. Its feathers glistened like the crystal-clear waters, and its wings were a masterpiece of nature's artistry. Yet, this extraordinary bird was renowned not only for its beauty but also for its extraordinary sense of direction. One day, however, it found itself lost.

That bright morning, as the sun painted the sky with hues of orange and pink, the confused seabird, exhausted from its endless flight, found itself in an unexpected place: a serene temple hidden deep in the heart of the kingdom's lush forest. Seeking refuge and rest, the bird gracefully landed within the temple's sacred grounds.

Word quickly spread among the temple's inhabitants about this magnificent and lost seabird. Soon, a curious crowd had gathered, marvelling at the creature's breathtaking presence. The temple's caretaker, a gentle soul who had dedicated his life to tending to the shrine, approached the bird with utmost reverence. He gently caught the bird and decided to present it to the king.

When the king received the magnificent bird, his eyes widened in awe. Never had he seen such a beautiful creature in his realm. He decided that a bird of such rare beauty deserved the finest treatment.

He ordered his chefs to prepare the best meat from the kingdom's most prized animals. He summoned the oldest and most exquisite wines from his vast cellars. And he commanded his musicians to entertain the bird with the most beautiful melodies ever played in his kingdom.

The bird, however, remained still and aloof, showing no interest in the lavish feast laid out before it. The king watched with growing frustration as the days passed, and the bird continued to refuse the tastiest meats, the aged wine, and the enchanting music. Three days passed, and the magnificent seabird, having neither eaten nor drunk a drop of wine, nor even acknowledging the music, finally succumbed to its hunger and exhaustion, dying within the opulent confines of the royal palace.

The king, struck by grief, lamented, "What a stubborn bird! I treated it just as I treat myself, with the finest food, the most aged wine, and the most beautiful music, and it still chose to die."

Meanwhile, a hushed silence had fallen over the palace. No one dared to tell the king that the bird did not eat meat, drink wine, or listen to music. The king, it seemed, had loved the bird to death, but he had loved it in the way he understood love – through lavish indulgence.

The seabird's tragic demise served as a poignant reminder to all who witnessed it. It underscored a profound truth that transcended the boundaries of that kingdom: happiness and love were not one-size-fits-all. What brought joy to one might not bring it to another. True love and care, the kind that nurtures and sustains, involved understanding and honouring the unique needs and preferences of the beloved.

And so, the memory of the magnificent seabird lived on, not just as a symbol of beauty, but as a lesson in empathy, reminding everyone that to truly love and care for another being, one must take the time to understand what brings them happiness and nourishment, for it is in that understanding that love can flourish and endure.

Misplaced Focus

In a beautiful countryside, there was a young lad named Arin who lived on a farm with his cute dog. Every day during the spring and summer the young boy would take his dog down to the pond on his farm and they would practice retrieving different items for hours. Arin was determined to be prepared for every situation that might arise when the duck season came.

Arin knew the people around him and that they are only looking for flaws in everything and everyone so that they can gossip. He resolved not to be a subject of their idle chatter. So, he had vigorous training day in and day out with his dog. A loyal and obedient companion, the dog diligently followed every command issued by the young boy.

The day of the duck season dawned, and Arin and his dog were the first to claim their spot by the pond, close to their farmhouse. The tranquillity of the morning was soon broken by the distant sound of ducks in flight. The boy slowly raised his shotgun, shooting three times at the flying herd with his third shot killing a duck, landing flat in the middle of the pond.

With an eager glance and a nod from Arin, the dog bounded through the tall grass and emerged at the water's edge. Instead of swimming in the water, as in practice sessions before, the dog walked on the surface of the water and astounded everyone. It picked up the duck and then trotted back to the boy.

Arin was left in awe, both amazed and bewildered by his dog's extraordinary ability to walk on water. However, he knew that no one would believe such a remarkable feat without witnessing it firsthand. Determined to validate his dog's astounding talent, he made a swift decision.

That very day, Arin paid a visit to his neighbouring farmer's house and extended an invitation for a joint hunting excursion the following morning. The farmer accepted eagerly, and Arin hurried home to ensure his dog was well-fed, trained, and rested for the day ahead.

On the appointed morning, Arin, his dog, and the neighbour set out, fully equipped for the hunt. They settled in the same duck blind by the pond, and once again, the sound of ducks in flight reached their ears. This time, Arin generously offered his neighbour the opportunity to take the shot. The farmer aimed his shotgun and took a couple of shots, successfully bringing down a duck. Following tradition, Arin granted his dog permission to retrieve the fallen duck.

Once more, the dog performed its astonishing feat, walking on water to fetch the duck. Arin, bursting with pride and astonishment, and finally, barely able to contain himself any longer turned to his neighbour and asked, "Did you notice anything extraordinary about my dog?"

The farmer, after a moment's contemplation, replied, "Indeed, I did observe something peculiar about your dog. Your dog can't swim!"

Arin was shocked that, despite seeing all the amazing things his dog could do, his neighbour was only interested in finding faults and did not believe what he saw. His neighbour's focus was on an unexpected detail.

In the end, Arin learned that sometimes, people see only what they want to see, missing the true marvels right before their eyes.

The Unconventional Justice

In the ancient city of Nablus, there lived a wise and respected Arab judge. He was known throughout the region for his fairness and wisdom in settling disputes. People from far and wide would come to seek his counsel.

One day a shopkeeper came to him to complain that goods from his shop were being stolen every night, but he was unable to catch the thief.

The judge listened carefully and then turned his attention to the door itself. The judge commanded the door of the shop to be taken off its hinges, carried to the marketplace and given 50 lashes because it had failed to do its duty of keeping the thief out of the store.

A huge crowd gathered to witness the strange sentence being carried out. When the beating was over, the judge stooped down and asked the door who the thief was. Then he put his ear to the door to hear better what it had to say.

Then he straightened himself and announced, "The door declares that the burglaries have been committed by a man who has a cobweb on the top of his turban."

Instantly, the hand of one man in the crowd went up to his turban to feel it. His house was searched, and the stolen goods recovered.

Triumph Over Trials

At the age of 15, Soichiro Honda embarked on a remarkable journey with a dream to work for Toyota, the largest engineering company of his time. With no formal education but unwavering determination, he left his modest Japanese hometown for the bustling streets of Tokyo.

However, Japan, like the rest of the world, was grappling with the Great Depression in the 1930s. Undeterred, young Honda, still a student, set up a small workshop where he tirelessly toiled on his innovative concept – the piston ring. He envisioned selling this groundbreaking idea to Toyota, but his initial attempts fell short of their standards.

He refused to give up. Rather than focus on his failure, he continued working towards his goal. To pursue his dream, Soichiro had to go back to school to have in-depth knowledge of the metals. He enrolled in a technical school with a single goal to figure out why he could not produce the piston ring at the required quality. After two years of redesigning, he finally came up with a piston ring that was of the quality Toyota asked for. After long days of making piston rings and long nights of studying, and a lot of adjustments in between, Soichiro seemed to have achieved his dream. After two more years of struggle, he won a contract with Toyota. They placed an order for 100,000 piston rings and they extended the capital to him to make his factory.

However, adversity persisted. The factory was bombed during World War II. Yet, Honda's spirit remained unbroken, as he declared, "I will build it up again." Facing a cement shortage, he ingeniously devised a new method using discarded gasoline cans from US fighters, which he fondly referred to as "Gifts from President Truman."

Dcspite enduring an earthquake that razed his factory, Honda refused to yield to defeat. As post-war Japan grappled with resource scarcity, gasoline was rationed and, in most cases, nearly impossible to find, and Mr. Honda could not even get enough gas to drive his car to the market to buy food for his family.

Yet again, instead of feeling helpless and defeated, Mr. Honda decided that he would not settle for this quality of life and started asking himself some very powerful questions: "How else can I feed my family? How can I use the things I already have to find a way to get there?"

He noticed a little motor he had, the size of a traditional lawn mower, and hooked it up to his bicycle. At that moment the first "motorbike" was created. He drove it to and from the market and it was not long before his friends and neighbours were asking him to make some for them too. Shortly thereafter he had made so many "motorbikes" that he ran out of motors, so he decided to build a new factory to manufacture his own.

Yet again another obstacle stood in Mr. Honda's way. He had no money and Japan was torn apart so how could he do it? Well, he came up with a brilliant idea to write to 18,000 bicycle shop owners in Japan, telling them that he had a cheap solution to getting Japan moving again and he wanted them to invest. Out of the 18,000 bicycle shop owners who received his letter 3,000 gave Honda the money to manufacture his first shipment. The "motorbike" was too big and bulky and very few people bought it. So, once again instead of giving up, Mr. Honda went to work on overcoming another obstacle. He changed his approach, and he stripped the motorbike down making it much lighter and smaller. He called it "The Cub" and it went on to become an "overnight success", winning Honda the Emperor's Award.

Everyone looked at Honda and thought how "lucky" he was to have come up with the idea, but luck had nothing to do with Honda's success. The key to Honda's success was how he reacted to all the problems and circumstances he encountered. Numerous times along his journey events were out of his control, but what led to Honda's success is what he could control. The people who succeeded also failed.

"Many people dream of success. I believe that success can be achieved only through repeated failure and self-analysis. Success is only one percent of your work, and the rest – bold overcoming of obstacles. If you are not afraid of them, success will come to you itself," Soichiro Honda.

The Lioness and the Young Jackal

Long ago, in a forest, Leo and Leonie, a contented lion couple, lived a peaceful life. Their existence was uneventful until the day Leonie gave birth to twin cubs. Leo, the lion, took on the role of the provider, hunting for his family.

One day, as Leo set out to hunt, he found himself empty-handed as the sun dipped below the horizon. With desperation nagging at him, he began his journey back home. It was during this journey that he came across a lone baby jackal. His heart softened upon seeing the helpless creature, and he couldn't bring himself to harm it. Instead, he gently carried the young jackal in his mouth and brought it home to Leonie as an unexpected gift.

Curious, Leonie asked, "My love, what is the meaning of this? Is this creature meant for our dinner?"

"It's a baby jackal I found on my way home. I couldn't bring myself to end its life; it's just a helpless infant. The choice is yours, my dear. You can decide its fate and choose whether to keep it or not," replied Leo.

"Since you spared its life, I will do the same. I'll care for this baby jackal as if he were my own. We shall name him Liam," declared the lioness.

And so, they chose to raise Liam as their own, feeding him with her milk. In time, Liam thrived, growing strong, and the three siblings played together harmoniously, unaware of their differences.

One day, while playing outside their den, they saw an elephant trumpeting nearby. Despite the enormous size of the elephant, the lion cubs instinctively took an attacking posture. However, Liam yelled, "Hold on! Don't go near that huge animal. He is our enemy."

He then ran back towards the den. It was the first time the cubs faced cowardice. They were disheartened to see Liam fleeing from the potential danger.

The cubs returned home gloomy and disappointed, recounting Liam's retreat to their mother, expressing their confusion about their brother's cowardice.

Leonie contemplated the situation deeply. She realized that it was time to reveal the truth about Liam's origin before his presence posed a danger to his own life. She also felt that since it was the time for the cubs to grow and develop their natural instincts, Liam's presence might hinder their growth.

With genuine concern, Leonie addressed Liam, "My son, the truth is that you are not a lion but a jackal by birth. I raised you as my own child when you were orphaned. You are smart, intelligent, and clever, but lions and jackals have different ways. I think it's best if you return to your own kind before our sons discover the truth and pose a threat to your wellbeing."

Upon hearing this revelation, Liam was overwhelmed with fear. Without hesitation, he fled from the den, disappearing into the depths of the jungle to reunite with his own kind, leaving behind the lion family that had once nurtured him.

The Legacy of Friendship

Bernard was a rich and accomplished businessman. All his life he worked so hard to provide a better future for his beloved family. His ten children, the joys of his life, were recipients of his immense love and care. Bernard was known to be a man of his word, and this quality had earned him abundant wealth and great friends.

Life went smoothly for Bernard and his family until his wife fell gravely ill and passed away. On her death bed, Bernard promised his wife to take care of their children and that he would give all ten of his children hundred gold coins each on the day of his death. He wanted their children to be well off and make a better living for themselves even after his death.

As the years passed, Bernard grew frail and weak, keenly aware that his children had not achieved the same level of success as he had enjoyed in his youth. At the time he made the promise to his wife, his wealth had exceeded the promised amount. However, in his latter days, he faced financial difficulties that cast a shadow on his ability to fulfil his commitment.

On the day of his death, all ten of his children assembled by their father's bedside, as he had requested. One by one, they approached their father, embraced him, and received a pouch containing hundred gold coins. Yet, when it was the youngest son's turn, Bernard gestured for everyone else to leave the room.

With a heavy heart, the ailing Bernard confessed to his youngest child, "I have terrible news. Though I have been able to give each of your brothers and sisters hundred gold coins, I have but only twenty for you."

"Father," the youngest son protested, "if you were aware of this, why did you not distribute the wealth more evenly among all your children?"

Bernard replied, "Son, it is always better to keep your word to as many as possible." He continued, "Though I cannot give you the wealth I promised, I offer you something far more valuable. In addition to the 20 coins, I have for you, I give you my ten dearest companions. Their friendship is worth more than all the gold I have ever possessed. I implore you to cherish them as I have."

The youngest son, though disappointed, had little choice but to accept his father's final wish.

Shortly after Bernard's passing, the nine older children, who were elated by their sudden windfall, embarked on extended vacations to enjoy their newfound wealth. Meanwhile, the youngest son remained at home, nursing his disappointment, and pondering the perceived injustice of his father's actions.

After settling his debts, he found himself with only four coins remaining. Reluctantly, he spent his remaining inheritance to host a dinner for his father's ten friends, as per his father's request.

To his surprise, Bernard's friends were deeply touched by the youngest son's gesture and the kindness he showed them. They rejoiced in the warmth of his hospitality and recognized the genuine affection he held for his late father's companions.

The following morning, each of the ten friends sent two cattle and a small sum of money as a token of their appreciation to the youngest son. Some of them even offered guidance on prudent investments. Before long, the youngest son had amassed a considerable fortune, surpassing the wealth of his nine siblings.

He became the wealthiest individual in the village, much like his father before him. Through this journey, he learned to appreciate the true value of good friendship. Above his desk, as a tribute to his father and a reminder of the priceless inheritance he had received, he inscribed these words: "Friendship is of more value than gold."

The Unwritten Success

In the busy streets of New York City, a poor peddler arrived from Lithuania, hoping for a better life. However, the streets of America did not gleam with the promised gold, and he soon resorted to selling humble wares like needles, pins, and hooks. Life was harsh, insults were abundant, and profits were meagre, prompting him to keep an eye out for better opportunities.

One day, he learned of a vacancy for a shammes, the caretaker of a synagogue on Attorney Street. Eagerly, he applied for the position.

The synagogue's president, a man of discernment, questioned him, "Can you read and write English?"

The peddler replied with regret, "No."

"Apologies, sir," responded the president. "In America, a shammes must be literate. You see, New York is not Lithuania."

With a heavy heart, the peddler departed, feeling the weight of missed opportunity. However, with time, his fortunes began to turn. He ventured into real estate and amassed a substantial fortune. One day, when he required a loan of a quarter of a million dollars to support a real estate venture, he approached his banker and made his request. To his delight, the loan was granted promptly.

"Feel free to write your own check," the bank's president offered warmly, extending a pen.

Embarrassed, the real estate figure stammered, "I'm sorry, I can't write at all. I've only learned to sign my name."

The banker's eyes widened with admiration, and he exclaimed, "Remarkable! If you've achieved such success without the ability to read or write, imagine what you could have accomplished if you did!"

With a wry smile, the realtor muttered, "Indeed, I would have been the shammes of the Attorney Street synagogue!"

From Milk Pail to Freedom

In a peaceful farmyard, a little frog leaped about in search of a delicious treat. He came across some crunchy flies but longed for something even more satisfying. His tiny adventures led him to the vicinity of the cowshed, where he decided to explore further.

With a quick hop, the frog entered the cowshed, chasing a mischievous fly. The fly, however, proved elusive, fluttering up to a milk pail before making its escape through an open window. The frog, in hot pursuit, leaped into the pail without a second thought.

Inside the pail, the frog found himself in a predicament. The milk level was too low for him to reach the top of the pail, and the sides were too high for him to climb out. Panic set in as he realized his predicament.

Desperately, the frog kicked and swam in frantic circles, trying to stay afloat. As exhaustion overcame him, he attempted to close his eyes for a brief rest, but that only caused him to sink deeper into the milk, nearly drowning.

Refusing to give in to fear and fatigue, the determined frog continued to kick and paddle with all his might. His legs grew weary, but he persevered. With each vigorous kick, something remarkable began to happen. The milk around him started to thicken and change its consistency.

Initially, the thickening milk made it even more challenging for the frog to swim and kick. It attempted to pull him down, but the frog remained undeterred.

His relentless efforts led to an extraordinary transformation. The milk slowly transformed into butter through the ceaseless churning and kicking of the frog. With resilience and unwavering determination, the frog's surroundings shifted.

Finally, the milk turned thick enough that the frog could stand atop it rather than sinking within. The milk had turned into butter, and the frog seized this opportunity. Climbing out to safety, he triumphantly returned to his family, forever remembered for his courage and persistence in the face of adversity.

The Lion without a Roar

In a jungle, a majestic lion named Sher reigned as the undisputed king. By his side was his cherished son, Aslan, whom he adored and hoped would succeed him as the ruler of the forest. Yet, Sher noticed something peculiar about his son – Aslan lacked the powerful roar that was natural to a lion.

Determined to help his son find his voice, Sher attempted to train Aslan to roar. Despite his best efforts, however, the young lion's attempts proved fruitless. His roars were feeble, and his father's hopes dwindled.

Seeking assistance, Sher turned to his friend, the wise elephant. Yet, all that the elephant could teach Aslan was to trumpet, quite different from the roar expected of a lion. Disappointed and frustrated, Sher questioned, "Is that a roar?" His anger led him to chase the well-meaning elephant away.

The next tutor was a fierce tiger, who displayed his fearsome teeth and attempted to teach Aslan to growl like him. Sher's disappointment deepened.

Finally, it was the clever fox's turn to instruct Aslan. The growling noises coming from his den shocked Sher. To his astonishment, he discovered his son growling alongside the fox. In a fit of frustration, Sher roared, causing the fox to flee. With that, Sher abandoned all hope of teaching his son to roar.

Aslan, recognizing his inability to roar at an early age, adapted in a unique way. He learned to treat all the animals with kindness and to truly listen to them. He honed the skill of persuading others without the need for a roar, winning the trust and affection of all the jungle's creatures.

One day, as Aslan played with his friends, the sound of gunfire echoed in the distance. Panic ensued among the animals as they fled, shouting, "hunter, hunter." But Aslan did not run. Before he could decide on a course of action, a bullet grazed his shoulder, and he let out an earth-shattering roar. The sheer force of his cry startled the hunter, who fled in terror.

Sher felt a surge of pride and happiness for his son,"At last, Aslan had found his roar!"

As Aslan grew older, he realized that he did not need to roar to command attention. His lack of a roar had made him an excellent communicator, able to engage in meaningful conversations and persuade others with his words. His legacy as the king who ruled not with a roar, but with conviction and compassion, endured in the hearts of all the animals in the jungle.

The Watered-Down Celebration

Once upon a time in a Nigerian village, a tribal chief named Omar sent out messengers with a special invitation to all the men in his tribe. They were invited to a grand feast, where the food would be provided, but there was a condition – each man had to bring one jug of palm wine.

Among the invitees was a young man named Kofi who eagerly wanted to attend the great festival, but he faced a dilemma. He didn't have any palm wine, and he was determined not to spend money on it. Instead, he hatched a crafty plan. He decided to save money by substituting water for the palm wine, thinking that in a large pot of wine, no one would notice if he added just one jug of water.

On the day of the feast, the village buzzed with excitement. The tribal drums began to beat early in the morning, reminding everyone of the upcoming celebration. All the men dressed in their finest clothes and gathered at the chief's house by midmorning. Laughter, music, and the mouthwatering aroma of food filled the air, creating an atmosphere of eager anticipation.

As each man arrived, he poured his jug of palm wine into a large earthen pot. Kofi, too, followed suit, carefully adding the contents of his jug, greeted the chief, and joined the dancers in the festivities.

When all the guests had gathered, the chief signalled for the music to stop and ordered the servants to fill everyone's glass with wine. The moment arrived for the grand toast. With a wide smile, the village chief raised his glass high and declared, "Let us all drink to the spirit of unity and friendship!"

As the chief spoke, all the guests raised their glasses and drank. However, a cry of disbelief suddenly echoed through the crowd, and they quickly took another sip. The chief exclaimed, "The wine is very tasty!" and to his delight, the crowd echoed, "Yes, the wine is tasty."

What they tasted was not wine, but water. Each guest had decided that his one jug of water could not spoil the great pot of palm wine.

As the festival continued, Kofi watched his fellow villagers dance and celebrate in blissful unity, realizing that honesty was the true key to savouring life's sweetest moments.

The Demon's Treasure

Long ago, a cunning demon guarded a vast treasure concealed beneath an ancient, dilapidated house. One day, the demon received an assignment that required him to journey to a distant corner of the world. So, he had to leave that area. He would not be able to return for twenty years. Contemplating the fate of his cherished treasure during his absence,the demon considered various options.

Should he employ a guardian to protect it? The thought crossed his mind, but he dismissed it swiftly, for such guardians would demand substantial compensation. Leaving the treasure buried under the old house was equally perilous, as it might tempt daring souls to unearth it and pilfer its riches. After much contemplation, a cunning idea dawned upon him.

The demon excavated the treasure and transported it to the house of a notorious miser. With a calculated smile, he addressed the miser, "Dear Sir, I am inclined to present you with a gift before I depart for a foreign land. I have always held you in high regard, and I implore you not to decline my offer. Feel free to expend these riches as you see fit. I have but one condition: upon your death, I shall inherit your entire estate."

The miser, driven by his greedy nature, readily accepted the demon's proposal. Grasping the treasure's potential, he saw this as an opportunity to amass even greater wealth.

Twenty years later, as the demon concluded his overseas mission and returned home, he received news of the miser's passing. He went to the miser's house, where he uncovered the treasure, still securely locked within its strongbox. To his amusement, not a single coin had been taken.

The demon chuckled at the irony of it all, realizing that the miser himself had unwittingly served as the most vigilant and cost-effective guardian of the treasure. The demon's cunning plan had proven successful, and he departed with his riches, knowing that even in his absence, the treasure had remained untouched, thanks to the miser's constant obsession with wealth.

As the demon vanished into the distance, he couldn't help but reflect on the curious twist of fate. In the end, he learned that sometimes, the very obsessions that imprison us can also become our unwitting protectors, guarding our most prized possessions when we least expect it.

The Threads of Guidance

In ancient China, there lived a renowned itinerant philosopher and sage named Mencius, a devoted follower of the great Confucius.

Mencius was born in the state of Zou in China. At an early age, his father died, so, he was brought up by his mother. In Chinese culture, Mencius' mother is celebrated as an ideal mother, an example of maternal devotion and wisdom, despite their modest circumstances.

Despite the family's poverty, she always ensured he had the best available education. She noticed that her son was strongly imitating the neighbours' behaviour, she realized the importance of the environment. Mencius' mother relocated their family three times in pursuit of a conducive atmosphere for her son's upbringing.

They eventually settled near a reputable public school, where diligent students inspired Mencius with their unwavering commitment to education and their courteous conduct. His young heart yearned to join them, and he implored his mother to grant him this opportunity. With satisfaction, she conceded, affirming, "This is the right path for my son."

Under the nurturing guidance of his mother, Mencius immersed himself in education, gaining knowledge and refining his character. As he grew older, his mother sent him to a prestigious school far from their home, where he resided in the school dormitory. However, the unfamiliar surroundings and homesickness began to weigh heavily on him. One day he sneaked out of the school and secretly returned home without informing the school authorities.

His unexpected return surprised his mother, who was at that moment diligently weaving at her loom. She looked up from the loom which she was weaving and asked him how he was doing at school. He answered her with an air of indifference that he was doing well enough. The wise mother did not scold her son, but picked up a pair of scissors and went back to the loom where she was weaving. Without a word, the mother suddenly cut the unfinished cloth on the loom in half.

"Why, Mother, did you destroy the unfinished cloth? I can see you've invested so much time and effort into it," a puzzled young Mencius inquired.

With tears glistening in her eyes, his mother gently explained, "My child, just as it takes time to weave a fine piece of cloth, it should take you time to receive an education in the best school we can afford. Yet, you have chosen to quit it halfway. What difference is there between my cutting this unfinished cloth and you abandoning your unfinished education?"

Mencius, perceptive beyond his years, comprehended the profound lesson his mother conveyed through her actions and analogy. He knelt before her and made a solemn promise to return to school promptly, vowing never to abandon his education until he had completed it.

The nurturing environment created by his mother's wisdom and love nurtured Mencius into one of the most esteemed thinkers in Chinese history, a testament to the power of parental guidance and the importance of perseverance in the pursuit of knowledge.

The Lightbulb

Once there was a young man, Mathews, who worked in a factory. His mentor,Venkat, an old technician, taught him to talk less, do more, and never stop developing his skills in every aspect of the factory's operation.

Ten years later, Venkat retired, and Mathews became a technician himself. He continued to do his work with the same dedication and diligence as he was taught.

One day, Mathews visited his mentor. The old man saw that he seemed unhappy, and asked what was troubling him.

Mathews sighed and poured his heart out: "I have been following your instructions exactly all these years. No matter what I work on, I keep quiet and focus on the job. I know I have done good work at the factory, and I have learned all the skills that can be learned there. What I don't understand is that the guys who don't have my experience or capabilities have all been promoted, while I am still making as little as I did before, when I was your apprentice."

Venkat asked, "Are you positive that you have become indispensable to the factory?"

Mathews nodded, "Yes."

The wise old man paced back and forth to think. After a while, he turned to Mathews, "You must request a day off, using whatever reason you like. It's time for you to give yourself a break."

The young man was surprised by this advice, but the more he thought about it, the more it made sense. He thanked his teacher and left quickly to make a time-off request.

When Mathews returned to work after his day off, the manager called him into the office to tell him that things did not go well at the factory while he was gone. Others encountered many problems that normally would be handled by him, and they had no idea how to solve them. Realizing his importance, the manager decided to promote him to the position of senior technician, to thank him and encourage him to keep up the good work.

Mathews was grateful for his mentor's wisdom. Surely, he thought, this was the secret to success!

From that point on, whenever the young man felt like he deserved more than what he was getting, he would take a day off. When he came back the next day, the situation would improve to his satisfaction.

This pattern continued for months. One day, Mathews found that he was blocked from going into the factory. Much to his shock, he found out that his employment was terminated. He could not believe it. Not knowing what else to do, he went back to see his mentor, to try to figure out how things had gone so wrong.

"Why did I lose my job?" Mathews asked with much wounded pride. "Did I not do everything as you instructed?"

"Actually, you did not, because you heard only half the lesson," Venkat shook his head. "You understood right away that no one pays any attention to a light bulb that is always on. It is only when it goes off that people suddenly take notice and realize they've been taking it for granted. You were so eager to apply this understanding that you left before hearing the second half."

"Second half?" it began to dawn on Mathews that perhaps he made a big mistake. "What was the second half?"

Venkat spoke slowly to make his point, "The second half, more important than the first, is the realization that if a lightbulb goes off frequently, then sooner or later it will be replaced with one that is more reliable. Who wants a lightbulb that no one can count on to provide illumination?"

"Nobody notices a lightbulb that is always on. Only the day it goes off is when people suddenly start noticing it", said the mentor. "However," Venkat said, stressing on the most important part of the lesson, "if a lightbulb goes off frequently, it is replaced by a more reliable one."

Golden Troubles

Once there lived a cobbler who radiated an air of happiness and contentment. Passersby would often smile and wave when they heard his joyful singing flowing from his shoe repair shop. Many found themselves drawn to his store not just for the fine craftsmanship but also to bask in the warmth of his ever-present smile.

Among those who observed the cobbler was a banker, a man of few songs and even fewer smiles. Restful nights eluded him, and the persistent good cheer of the cobbler initially irritated him. However, as days turned into weeks, he found himself mysteriously drawn to the cobbler's shop and decided to find out the secret behind the cobbler's happiness.

After a pleasant conversation, the banker couldn't help but ask, "Are you a wealthy man? Forgive my curiosity, but could you tell me about your annual income?"

"My family is seldom in want," replied the cobbler. "There are days when my shop sees no customers, and on holidays, we close entirely, with no income at all during the festive season. It's hard to provide an exact figure."

"Remarkably uncomplicated," mused the banker. "Considering how openly you've shared your life with me, I've decided to ease your financial concerns in the short term. Please accept this gift of 300 gold coins, to be used as needed."

Overjoyed, the cobbler hastened home and carefully concealed the gold in a corner of his house. However, as days passed, his once-tranquil life began to change. He frequently left his shop to return home when his family was absent, fearing that someone might seize his newfound wealth when they were away. Night after night, sleep eluded him as he worried about potential plots to steal the gold. Old friends noticed that his songs lacked their former cheer, and he became increasingly suspicious when customers visited the shop merely for a friendly chat.

Eventually, the cobbler visited the banker, clutching a bag of gold coins in his trembling hands. He began, "Thank you for your generous gift," he began, "but I cannot really afford to be the owner of these gold coins. Please take your money back so that I may once again embrace music, enjoy peaceful sleep, and cherish the company of my friends. It seems that when I buried the money, I buried my happiness with it."

The cobbler realised that wealth could bring comfort. But it could also burden us with worries and overshadow the simpler, more precious moments that bring true happiness. True happiness often lies in contentment with what we have, rather than the pursuit of excessive riches.

What the Little Pig Missed!

In a village near the little town of Bethlehem, there lived a community of animals who were known for their extraordinary kindness and compassion. This quaint village was tucked away from the bustling world, and the animals here had formed a unique bond with the humans who resided nearby. It was rumoured that a special event was to take place on the night of December 24th.

It was about midnight on December 24th. Close to Bethlehem, the shepherds kept watch over their flocks. Suddenly, the Angel of the Lord appeared to them and joyfully sang, "Rejoice! I bring good news for all the people. In your own town, in Bethlehem, your Saviour is born. You will find him in swaddling clothes lying in the manger in that cave down the mountain." Immediately, the shepherds left the sheep behind and rushed to the cave. Whilst the shepherds were away, all the animals gathered.

Thus, they said, "Our shepherds are gone to see the Child. Yeah, what about us? I want to go too! Yeah, let us. We all want to go. Come on."

And they rushed down the slope towards the cave.

When they reached the cave, they pushed and elbowed one another just to get in first.

Then, a hush fell on them all. The sight was too much for them. No one moved or made any noise.

They were spellbound by the charming, bright, smiling Little Child. Only the little pig did not seem to be very much taken up by the newborn. He remained alone in a corner of the cave, nosing about for some scraps of food.

After a long time, the animals left the cave. All were silent, full of awe and admiration.

After a while they began talking and commenting on what they had seen and felt:

"What a lovely child!

"What bright and sparkling eyes!"

"What plump and rosy cheeks!"

"What a charming smile!"

All wanted to say something. They were so eager. Only the little pig remained silent. He said nothing, kept on nosing about the ground. At last, the little pig growled:

"What's all this fuss about? Eh? Child? Sparkling eyes? Plumpy cheeks? Pooh! I saw nothing at all! All I saw was mud, straw, cow dung."

A little goat retorted, "Of course, you are a pig! Dung is all you look for. Dung is all you see. Dung is all you get."

Full of joy and praising God all the animals went back to their flocks. The little pig, sad and dejected, followed them at a distance. What joy could he have? What could he be singing about? He could not figure out why he did not see the Child, the sparkling eyes, the plump cheeks, and the charming smile!

The Vineyard of Generosity

Abdullah Jaafar, the proud owner of a flourishing orchard, embarked on a visit to his cherished estate one day. As he strolled through the serene landscape, he passed by a vineyard and noticed an unusual sight. A black slave sat amidst the vines, a few pieces of bread laid out in front of him, and a hungry dog keeping him company.

Intrigued, Abdullah observed as the slave compassionately tore a piece of bread and tossed it to the famished dog. He repeated this act of kindness, sharing his meagre meal with the canine companion.

Curious, Abdullah approached the slave and inquired, "How much bread do you receive each day?"

The slave, with a serene countenance, replied, "The exact quantity you just witnessed, sir, which is now shared with this hungry dog."

Puzzled, Abdullah questioned, "But why do you choose to feed a dog before attending to your own hunger?"

The slave explained, "There are no other dogs around here, sir. This one has travelled from afar and is in dire need. I couldn't bear to consume my bread while knowing this loyal creature was hungry."

A sense of admiration welled up in Abdullah's heart as he marvelled at the slave's selflessness. He couldn't help but ask, "But how will you manage to sustain yourself today without food?"

The slave responded with determination, "I will endure the hunger, sir."

Deeply moved, Abdullah contemplated, "I am known for my reputation of generosity, and yet, this humble slave has exhibited a generosity that surpasses my own."

In a spontaneous act of compassion and benevolence, Abdullah decided to purchase the vineyard and gifted it to the slave. Not stopping there, he also secured the slave's freedom, granting him a new lease on life.

In the end, it was not just the vineyard that had changed hands; it was the vineyard of generosity itself, sown in the heart of Abdullah, that had truly blossomed.

The Ambitious Mosquitoes

Once upon a time in the animal kingdom, there lived a bold mosquito with big dreams. She was determined to defeat the mightiest creature of all, the horse. So, she approached the horse with a daring question, "If someone could beat you, would he be considered the greatest?"

The horse, confident in his strength, replied, "Certainly, he would be. But no one can defeat me."

Undaunted, the mosquito made a daring attempt to sting the horse, but her efforts were in vain. She felt disheartened and decided to seek the counsel of the wise hare, known for his wisdom.

When the mosquito told the hare about her encounter with the horse and her unfulfilled ambitions, the hare pondered for a moment before offering his sage advice. He said, "Gather all your fellow mosquitoes, unite your efforts, and sting the horse relentlessly. Together, you can achieve your goal."

Inspired by the hare's wisdom, the mosquito rallied her mosquito companions. Following the hare's guidance, they set out together, located the horse, and tirelessly stung him until he died.

The mosquitoes celebrated their remarkable success and returned to the wise hare, eager to share their triumph. Word of their victory spread, and animals from near and far sought the hare's counsel, hoping to be part of his followers.

However, amidst the jubilation, no one inquired about the actual accomplishment. Each mosquito saw himself as the hero. They scattered in all directions, attempting to sting various creatures. Yet, their stings repeatedly proved ineffective in causing the death of any animal.

Occasionally, a swarm of mosquitoes did manage to harm an unfortunate creature, fuelling the belief that stinging held great power. In some circles, the mosquito was regarded as a formidable force, second only to the wise hare who had provided the initial guidance. Yet, the true significance of their actions remained shrouded in mystery, lost to the ages.

One needs to look beyond the surface and acknowledge that the impact of an action can be more complex than it first appears.

The Cunning Kingship

In the heart of a jungle, a clever fox named Rubah found himself in a tight spot when he encountered a hungry tiger named Macan. Macan was ready to make a meal out of Rubah, who had to think fast.

Rather than showing fear, Rubah decided to take a different approach. He maintained a calm and serious demeanour, addressing Macan with authority. "Hold on a moment, Mr. Macan. You cannot harm me. I've been chosen by the heavens as the king of this jungle. Every creature here must show me respect. Harming me would go against divine will, you understand?"

Macan, surprised by this unexpected proclamation, released Rubah saying, "I've never heard of such a thing."

Seizing the opportunity to give proof of his newfound status, Rubah proposed, "Let's put my claims to the test, Mr. Macan. We'll venture into the forest together, and you will witness how the animals cringe before me in reverence." Macan, still sceptical but intrigued, reluctantly agreed.

As they walked, Macan warned, "I'll chew every bone of you if you're lying." To Macan's surprise, every small animal they encountered fled in terror upon seeing Rubah with the mighty tiger. The evidence spoke for itself, and Macan had to acknowledge the truth of Rubah's claims.

Rubah, feeling triumphant, asked, "Is there any doubt now?"

Macan bowed respectfully, admitting, "Your Highness, you are indeed the rightful king of these woods." With that, Macan withdrew, and Rubah, feeling secure, climbed a tall tree for a relaxing nap, knowing no formidable beast could reach him there.

As Rubah reclined in the lofty branches, he shared a knowing glance with a wise old owl perched nearby. The jungle seemed to whisper its approval, acknowledging the irony of a cunning kingship that brought harmony to its wild domains.

"Rubah" is the Indonesian word for "fox"

"Macan" means "tiger" in Indonesian.

Beneath the Featherbed

Emperor Napoleon Bonaparte, a prominent military and political figure in the late 18th and early 19th centuries, rose to prominence during the French Revolution. He became Emperor of the French in 1804 and, through his military conquests, and expanded the French Empire. However, his ambitions faced setbacks, including the ill-fated invasion of Russia in 1812.

During the infamous retreat from Russia, Emperor Napoleon found himself in extreme difficulties as he fled from the pursuing Russian forces. The unforgiving winter and relentless enemy pursuit left him with few options. In his desperate flight, he stumbled upon a humble Jewish village, seeking refuge from the approaching danger.

With the Russians hot on his heels, Napoleon's plea was frantic as he implored the first person he encountered, a Jewish tailor, "Hide me, please! If the Russians capture me, it will mean my end!"

Though the Jewish tailor had no inkling of the stranger's identity, his heart was moved by compassion for a fellow human being. He swiftly directed the emperor, "Quick, get beneath the feather bed!"

Napoleon obeyed, taking refuge beneath layers of feather beds piled one upon another. It wasn't long before the door of the tailor's home burst open, revealing two Russian soldiers brandishing spears. Their harsh question filled the room, "Is anyone hiding here?"

The tailor, quick on his feet, replied calmly, "Who would be foolish enough to hide in my humble abode?" The soldiers, thorough in their search, scoured every nook and cranny but found no trace of the emperor. To ensure their thoroughness, they thrust their spears several times into the feather beds.

As the soldiers departed, satisfied that their quarry was not present, Napoleon, pale and drenched in sweat, emerged from his concealed refuge. Grateful for his escape, he turned to the tailor and declared, "You have saved my life, my noble friend. In gratitude, you may ask three favours, and I shall grant them without hesitation."

The humble tailor pondered for a moment, then made his first request, "Your Majesty, the roof of my house has been leaking for two long years, but I lacked the means to repair it. Could you have it fixed for me?"

Napoleon, somewhat exasperated by this seemingly trivial request, granted it nonetheless, saying, "Very well, I shall see to it that your roof is repaired. Now, for your second request, please choose something more substantial."

The tailor contemplated once more and then brightened, "Arival tailor opened a shop across from mine, and he is ruining my livelihood. Could you kindly ask him to relocate elsewhere?"

Napoleon, somewhat frustrated but bound by his promise, agreed, "Very well, I shall ask your competitor to find another location. But this is your final request."

The tailor, a mischievous glint in his eyes, asked, "Emperor, may I know how you felt when the Russian soldiers thrust their spears into the feather beds?"

Napoleon, incensed by the question, exclaimed, "You impudent fool! How dare you pose such a question to an Emperor! For your insolence, I shall have you executed at dawn!" The tailor was immediately placed in irons and led away to a guardhouse.

Throughout the night, the tailor could not sleep, wrenched with fear and sorrow. He prayed and sought forgiveness from God for his impending fate.

At dawn, the tailor was led to a tree, where a firing squad stood at the ready, muskets aimed at him. An officer with a watch in hand prepared to give the order to fire. As he began the countdown, "One, two, three..."

Before the word "three" was even complete, a commotion erupted as the emperor's aide-de-camp galloped up on horseback, urgently shouting, "Halt! Do not fire!"

Then he went up to the tailor and said to him, "His Majesty, the Emperor, gives you his gracious pardon. He also has asked me to give you this letter."

The aide-de-camp approached the trembling tailor, delivering the emperor's pardon and handing him a letter.

With a profound sense of relief, the tailor read the letter, which contained a simple message from Napoleon, "You wanted to know, how I felt under the featherbed in your house. Well, now you know!"

The Generous Lion and the Crippled Fox

Once upon a time, there was a man who yearned for divine guidance in his quest for enduring happiness and success. After months of fervent prayers, he had a significant dream. In the dream, he wandered in a forest, searching for enlightenment. Taking this as a sign from a higher power, he ventured into the woods the following day, seeking answers.

During his journey, he encountered an unusual sight: a legless fox, comfortably resting between two rocks in a shaded spot. Despite its obvious disability, the fox seemed well-fed and healthy, which was quite remarkable given nature's usual indifference to disabilities. Intrigued, the man hid behind a tree and observed the fox until sunset.

To his astonishment, a lion appeared and shared a portion of its meal with the legless fox. The fox enjoyed its meal and slept peacefully. He could not trust his eyes. It felt like a miracle, a message from God.

Overwhelmed by this revelation, the man thought, "Ah, I have finally grasped the truth. Success in life lies in trusting that God will provide for all my needs. There's no need for me to toil and struggle. I must simply surrender and have faith."

However, after enduring two challenging weeks of starvation and weakness, the man received another dream. In this dream, he heard a voice saying, "Fool! Be like the lion, not like the fox."

Upon waking, the man had a profound realization. He understood that the lion, with its diligence and self-reliance, could not only sustain itself but also extend its help to others. On the contrary, the legless fox, seemingly dependent on divine favour, had thrived largely due to the lion's generosity. The lesson was crystal clear: while faith and surrender had their place, they needed to work hand in hand with determination and action, much like the lion's unwavering self-sufficiency.

The Heart of a King

In a distant island kingdom, a king lay ill, pondering over the question of who would be his worthy successor. With no heir of his own, he yearned to appoint someone with a genuine love for their fellow human beings. There were thousands of young men in his kingdom aspiring to become the king!

The king issued an edict throughout the land, inviting any young man who believed he had the heart of a king to step forward and meet him, "Any young person who thinks he has a claim to be the next king, come forward and meet His Majesty. The only qualification for the candidate was that he should have immense love for his fellow human beings."

Every young man across the country was very excited at the prospect of being the ruler of the country.

Among the eager aspirants was a humble youth named Henry, from a remote village who also read the notice and started to prepare for an interview with the king. The man was kind and hard-working but was very poor. He had no good clothes to present himself before the king. Undeterred, Henry toiled tirelessly, saving every penny to purchase suitable clothing and provisions for the journey.

So, the young man worked hard and saved some money to buy him proper clothes, shoes and supplies to support him for the long journey to the palace. After he got his supplies and proper clothes for the interview the young man set out on his quest.

He travelled many days and had almost completed the journey. When he came upon a poor beggar by the side of the road, the poor beggar said trembling in the cold, covered only in tattered rags.

Henry was moved to pity and asked him, "What can I do for you my dear man?" The beggar replied, "Please help me. I have no clothes, no shoes to wear and it is so cold. Do something for me."

Henry said, "I have no other clothes and shoes than these I wear, yet if you really wish to have them, let us exchange them." The young man was so moved by the pathetic condition of the beggar that he immediately stripped his new attire and offered it to him. He also shared with him the very limited supply of food he had carried for the journey.

Henry was a bit hesitant to go for the interview. He, however, gathered enough courage to enter the palace in his old, dirty clothes. Upon his arrival at the palace, the king's attendant showed him the way to the Great Hall. The man was admitted before the throne for the interview.

With humility, he bowed as the king entered the room. Yet, when he raised his gaze, astonishment overcame him. The king bore a striking resemblance to the beggar he had encountered on the road.

Observing the shock in Henry's eyes, the king calmly acknowledged, "Indeed, I was the beggar you met on your journey."

Perplexed, Henry stammered, "But why did you disguise yourself as a beggar when you are the king?"

The king replied with wisdom in his voice, "I needed to make sure that the new king possessed a noble heart, one filled with genuine love for fellow human beings. Had I revealed my true identity, you might have done anything to impress me, I would never have known what is truly in your heart."

He continued, "Generosity and compassion, given freely to those in need without expecting anything in return, are the signs of a great heart. Your love and generosity towards the beggar have revealed your sincere care for others. This nation requires a leader who will work tirelessly for the welfare of the entire population, not just those seeking the throne."

With a smile, the King declared, "You have proven yourself to be my rightful successor. Your heart is worthy of the throne, and you shall lead this kingdom with compassion and love."

Chatter Lessons

In a small village, a classroom was filled with eager and curious students. Among them, there was a boy named Mohan, known for his incessant chatter. He had an endless supply of questions and stories to share, making him a beloved friend to his classmates. However, there was one place where his talkative nature sometimes got him into trouble - the classroom.

Whenever the teacher posed a simple question, Mohan would launch into lengthy explanations, often accompanied by animated gestures and a tone of excitement. The teacher advised him to be concise and get to the point, but the boy couldn't understand why.

One day, the teacher asked, "Can you tell me which creatures make a lot of noise yet go unnoticed?"

Mohan thought for a moment and replied, "Crickets and toads."

"Correct," the teacher nodded. "Now, can you tell me which animal everyone pays attention to when it makes sounds?"

Mohan answered, "Certainly, it's the rooster. Its crowing wakes people up every morning."

The teacher then posed a final question, "So, would you prefer to chatter endlessly like crickets or toads, being ignored, or would you rather speak like the rooster, commanding attention with your words?"

Mohan got the point.

The Miraculous Cure

Long ago, there reigned a king who was very lazy. His days were spent in idleness, with sleep and food being his only companions. As one might expect, this idle lifestyle eventually took a toll on his health, and the king fell ill.

The royal physician, though aware of the true cause of the king's ailment, dared not utter a word. Many doctors were summoned from all corners of the kingdom, but they all shook their heads in despair, declaring, "This illness is unlike any we have ever seen. Alas! There is no cure to be found."

Desperate for a solution, the king was told of a wise hermit known for his ability to treat the most peculiar of maladies. Without delay, he sent for the hermit.

Upon hearing of the king's condition, the sage nodded knowingly. "Your Majesty, I can indeed help you, but the path to recovery requires effort on your part. You must come to my hermitage every morning for the treatment of your ailment."

The hermit handed the king a pair of substantial dumbbells. "Hold these magical spheres in your hands and swing them for half an hour every morning before your breakfast. If you faithfully perform this exercise, your health shall be restored. However, cease this routine, and your illness will surely return."

The king, willing to do whatever it took to regain his health, committed himself to the hermit's regimen. Each day, he embarked on the journey to the hermitage, engaging in his newfound exercise with the dumbbells.

Weeks passed, and the transformation was undeniable. The once-lazy king shed his excess weight, and his health was fully restored. The miraculous cure had come not from potions or physicians but from the king's own determination.

The Unchangeable Nature

In a lonely desert, on a faraway mountain, there was a scorpion named Ted. He had made this barren land his home for as long as he could remember, residing within the confines of a dim and desolate cave. The cave was a respite from the scorching desert heat and relentless winds, providing a semblance of shelter amidst the harsh terrain. His days were marked by slumber, nestled between the cool rocks, while the nights saw him burrowing beneath the warm sand.

As time passed, Ted found himself growing increasingly weary of his solitary existence. The monotony of his life had become stifling, and a restless desire to explore the world stirred within him.

One day, Ted's journey led him to a different landscape—a grassy, damp riverbank, a stark contrast to the arid desert. The river before him flowed wide, deep, and swift, presenting a formidable obstacle to his path. The sun was beginning its descent, leaving Ted with little choice but to find a way across the river.

Pausing to contemplate his predicament, Ted's eyes fell upon a laid-back frog perched on a moss-covered stone on the opposite bank. The frog seemed to be simply observing the world go by.

Summoning his courage, Ted called out to the frog across the expanse of water, "Hello, Mr. Frog! I am Ted, hailing from the desert mountains. It's a pleasure to make your acquaintance."

Startled by Ted's unexpected presence, the frog, named Green, leaped closer to the river's edge, perching himself upon a nearby stone. His cautious demeanour was evident in his response, "Hello, Ted. I am Green. What brings you here, and what can I do for you?"

Ted explained his situation to Green, conveying his desire for assistance in crossing the river. He continued, "I'm new to this area, and I've never encountered such a river. Unfortunately, I cannot swim, and the setting sun leaves me with limited options. Would you be so kind as to offer me a ride on your back across the river?"

Green, however, remained cautious. He replied, "It's nearly nightfall, and I have reservations about helping you. How can I be sure you won't harm me? I've always been told to be wary of scorpions. Why should I take the risk?"

Ted persisted, explaining his vulnerability, "You see, I'm unfamiliar with this river, and I have no knowledge of swimming. If I were to sting you, we would both perish."

After careful consideration, Green reluctantly agreed to help Ted. He swam to the riverbank and positioned himself near the muddy shore, ready to carry Ted across the river. He clambered onto Green's back, his sharp claws prickling against the frog's delicate skin as they embarked on their river crossing.

The water surged around them, but Green swam skilfully, keeping Ted near the surface to ensure he did not drown. With strong and purposeful kicks, Green propelled them through the first half of the river, his flippers working tirelessly against the powerful current.

However, as they reached the middle of the river, where the current was strongest, the frog suddenly felt a sharp, excruciating pain in his back. Out of the corner of his eye, he witnessed Ted withdrawing his stinger. A numbing sensation began to creep through his limbs.

Green, paralysed by fear, cried out in a croaking voice, "You fool! Why would you do such a thing? Now, both of us are doomed! Why did you sting me after my trust?"

Ted, gripped by remorse, as they began to sink beneath the water, replied with a sad inevitability, "I'm sorry, Mr. Green. It's in my nature to sting. I couldn't help it!"

Ted shrugged and did a little jig on the frog's back. Then they both disappeared into the darkness of the muddy waters of the swiftly flowing river.

The Bird in the Golden Cage

Once upon a time, there lived a merchant who possessed a remarkable bird. This bird, with splendid plumage, was kept in a golden cage. The merchant, planning a journey to India, inquired of the bird if it desired anything from its homeland. The bird yearned for its freedom, a plea that was met with a cold refusal. Instead, it asked the merchant to travel to an Indian jungle and share its captivity with the free birds dwelling there.

The merchant, being a kind-hearted soul, fulfilled the bird's request. As the man spoke of the captive bird to the wild ones, a bird that resembled his own fell from a tree in a dead faint.

Upset, the merchant thought, "I had no idea that birds could be so sensitive."

Convinced that this bird must be a relative of the one in his golden cage, he returned home with a heavy heart. When the captive bird inquired about the news from India, the merchant sadly replied, "I'm afraid I bring you bad tidings. One of your kin collapsed at my feet when I mentioned your captivity."

Upon hearing these words, the captive bird slumped to the bottom of the cage, seemingly lifeless.

The merchant was filled with regret, thinking, "I never intended my words to be so powerful."

In an act of mourning, he opened the cage and gently placed the bird on the windowsill, preparing to bury it. As he stepped out of the room, the bird miraculously revived and exclaimed, "I knew my relatives would offer wise counsel."

With a smile, the bird continued, "You see, what appeared to be a disaster was actually good news for me. The message and the suggestion on how to free myself were transmitted through you, my captor."

With that discernment, the bird took flight and soared away, finally liberated from its golden cage.

Running for Life

There lived a famished cheetah in a forest, its stomach growling with hunger. One day, it spotted a fleet-footed rabbit darting through the grasslands. The cheetah's instincts took over, and it embarked on a relentless pursuit that lasted for three exhausting hours. Yet, despite its unmatched speed, the cheetah found itself unable to close the gap and capture the elusive rabbit.

Breathless and desperate, the cheetah decided to employ a different strategy. It called out to the rabbit, desperation in its voice, "My dear rabbit, please wait for just a moment. I give you, my word; I won't harm you. I simply wish to ask you a question."

The rabbit, cautious and wary of the cheetah's intentions, replied cautiously, "I'll wait for a minute, but you must stay at a distance. Go ahead and ask your question."

With a sigh of relief, the cheetah inquired, "I am renowned as the fastest runner in the animal kingdom. I've been chasing you for three long hours. How is it possible that I haven't been able to catch you?"

The rabbit, still vigilant, replied with a knowing smile, "Well, my dear cheetah, the answer is quite simple. You see, you are running for your next meal, but I am running for my very life. That's why, despite your incredible speed, you could not catch me."

With that profound truth, the rabbit continued its escape, disappearing into the wilderness, leaving the cheetah to ponder the wisdom of the rabbit's words. In the wild, survival often depended not just on swiftness but also on the unyielding determination to stay one step ahead of those who sought to make you, their prey.

Priceless Discovery

In a quiet riverside village, there lived a humble fisherman named Raju. Every morning, before sunrise Raju would set out on his weathered boat to cast his net into the flowing waters of the river. His daily ritual had always been to wait patiently for the first rays of dawn before beginning his work, as it was only then that the river would reveal its treasures.

One morning, as the world was still covered in darkness and the river was shrouded in mist, Raju decided to wait a little longer before casting his net. He sat down on a smooth boulder by the riverside, dipping his feet into the cool, flowing water. As he sat there, he felt something unusual beneath his feet.

Curiosity excited, he reached down and retrieved a cloth bundle from beneath the boulder. His fingers traced the shape of the bundle, and he realized it contained something. With a sense of curiosity, he slowly ripped out the cloth, revealing a small bag. It felt heavy, and inside it, he could feel the smooth, round surfaces of what he assumed were river pebbles.

Fascinated by the contents of the bag, Raju decided to investigate further. He took one of the round objects in his hand and, without much thought, tossed it into the river. As the pebble hit the water, it created a delightful "plop" sound that resonated with a satisfying harmony. The sound was enchanting to his ears, so he decided to repeat the action with another pebble, relishing the musical "plop" it produced. He repeated this process with twenty two more pebbles, each time enjoying the soothing sound they made as they met the water's surface.

Finally, Raju was left with just one pebble in his hand. He decided to examine it more closely now that the sun had risen, casting its gentle light on his surroundings. When he looked at the last pebble, he was taken aback. It was not a pebble at all but a priceless jewel, gleaming brilliantly in the morning light.

Overwhelmed by a mix of emotions, Raju exclaimed, “What! God blessed me with twenty four of these precious gems, and in my ignorance, I tossed away twenty three of them into the river! How unfortunate I have been!”

Regret gnawed at his heart as he realized the true value of what he had carelessly discarded. From that day forward, Raju vowed never to let ignorance cloud his judgment and to always cherish the treasures, whether they be material or moments, that life presented to him.

The Deaf Frog

Once in a forest, a group of frogs embarked on a journey. Along their path, two of these frogs accidentally stumbled into a deep and treacherous pit. As their fellow frogs gathered around the pit, they peered down and realized the gravity of the situation. Their voices were filled with despair as they declared, "There's no hope left for you."

However, the two frogs trapped in the pit had a different perspective. They chose to disregard the bleak pronouncements of their companions and decided to take matters into their own hands. With determined hearts, they began to leap and jump to escape the pit's grasp.

Despite their best efforts, the group of frogs watching from above continued to offer discouragement. They shouted, "Give up! You'll never make it out." The chorus of voices echoed through the pit as they proclaimed, "Hold on, you frog! Your strength is insufficient to scale this deep pit. A fall from such a height will surely cost you your life. It's wiser to refrain from trying."

Undeterred, one of the frogs paid heed to the cautionary words of the others and gave up, succumbing to the pit's depth and meeting an unfortunate end. Meanwhile, the other frog persisted in its struggle, jumping with all its might. Once more, the frog's companions shouted, urging it to end its suffering and accept the inevitable.

The frog, now halfway to freedom, was exhausted but resolute. It leaped with increasing determination, driven by an indomitable will. Finally, after a strenuous effort, it emerged from the pit. Gasping for breath and weary but triumphant, the frog stood on solid ground.

Upon seeing their fellow frog's success, the others above the pit exclaimed, "Did you not hear us?" The frog, panting but victorious, explained that it was deaf and had perceived their shouts as cheers of encouragement. The frog had thought the cacophony of voices was a testament to their unwavering support and belief in its ability to overcome the pit's challenge.

The other frogs, realizing their unwitting role in the deaf frog's escape, rejoiced, and said, "We are glad that you made it out, despite what we screamed." This tale served as a reminder that sometimes, in the face of adversity, unwavering determination and the belief in oneself can lead to success, even when the world seems to be shouting otherwise.

A Never-Ending Story

Akbar was a Mughal emperor who lived in India in the sixteenth century. Birbal was the cleverest of all his ministers. He was also a man of ready wit and had a good sense of humour.

Once when Emperor Akbar fell ill, it prevented him from sleeping till dawn. The royal doctors tried many medicines, but they did not work.

Then a hakim suggested that the king should hear a story before retiring to bed every night so that his mind might remain calm and cheerful and help him to fall asleep.

"It seems a good suggestion," said the king. "But who will tell me a story every night?" he asked.

"That is easy, you have so many courtiers. Each one of them in turn can tell you a story," suggested the hakim.

He accepted the suggestion and ordered his courtiers to tell him a story every night in turn. It became a problem for courtiers. Though they managed to tell a story, the emperor would always ask, "What next?" So, the storyteller had to tell another story. This went on till the storyteller got tired and was at his wit's end.

So, the courtiers approached Birbal and asked him to find a solution to this problem. Birbal asked them not to worry and he took it upon himself to go to the emperor that night to tell him a story.

That night the emperor welcomed Birbal happily. Birbal narrated the story to him. At the end of the story the emperor asked, "What next?" as it was his habit. Birbal told him another story. At the end of the story Akbar again asked, "What next?"

To put an end to this habit, Birbal started another story. It went like this. "There lived a farmer in a small village. He worked very hard to produce the crops. Unfortunately, a major part of the harvest was consumed by the sparrows. The farmer got desperate. He decided to solve this problem."

"What happened next?" asked the emperor.

"And then…",Birbal continued the story. "The farmer built a big barn to store the food grains. It was made of strong bricks and cement. It has large wooden doors. Not even a worm would enter it. Now the sparrows were in trouble. They did not know what to do." said Birbal.

"And then…," Akbar asked again.

Then the sparrows called an emergency meeting. They started discussing their problem. One little sparrow said, "This stupid farmer has locked the food grains in the barn. How dare he do that?"

"Calm down, dear," said an old sparrow in a feeble voice, "I have an idea. We can hire a mouse who would make a hole in the door of the barn. Then, we will be able to satisfy our hunger."

All the sparrows approved of the idea. They hired a mouse. After some negotiations, they decided on a suitable remuneration for him. "And then…" said Akbar, again.

Then, the sparrows took the mouse to the barn. The mouse closely inspected the doors and started his work. Within no time, he made a hole in the door.

"What next?" said the emperor.

"Sir, then the first sparrow went inside the barn, satisfied her hunger and returned," said Birbal.

"And then," Akbar said again.

"Then the second sparrow went inside the barn, satisfied her hunger and returned," continued Birbal.

"And then," asked Akbar.

"Sir, do you know what happened next?" said Birbal in a very excited mood. Akbar opened his eyes.

"Sir, then the third sparrow went inside the barn, ate the food grains and came outside," Birbal repeated, "And then the fourth sparrow…."

"Stop it, Birbal," said Akbar, annoyed. "What are you doing? Why is the story not proceeding further?"

"Sir, there were hundreds of thousands of hungry sparrows. The story cannot move further before they all have satisfied their hunger," replied Birbal.

Then the Emperor got angry and shouted, "Birbal when are these birds going to finish eating the grains?"

"When you stop asking "What next," said Birbal.

The emperor realized his folly. He ordered the program of storytelling to be stopped. By this time, he had also regained his health.

Running for Health

It was in the days of Czar Nicholas II, the last Czar of Russia before its revolution in 1917. He issued several antisemitic laws which led to mass emigration of Jews from Russia. Those who chose to remain in Russia were driven or persecuted. Joshua and Amos, two Jews, were walking along a main road in Moscow. Joshua had a residence permit, Amos did not. Suddenly a policeman appeared.

"Quick, run!" whispered Amos, the one without the permit. "When the policeman sees you run, he will think you have no permit, so he will run after you. This will give me a chance to get away, and it won't hurt you any because you can show him your permit."

So, Joshua, the Jew with the permit started to run. As soon as the policeman saw him do so, he went in hot pursuit. After a few minutes he caught up with the Jew.

"Aha!" gloated the policeman. "So, you have no permit!"

"No permit! What makes you think I have no permit?" asked Joshua, showing it to him.

The policeman looked bewildered.

"Why did you run away when you saw me?"

"My doctor advised me to run five miles a day for my physical fitness."

"But didn't you see me running after you?" asked the policeman.

"Sure, I did. But I thought your doctor had given you the same advice."

The Secret of Power

In ancient times, a mighty turbulence stirred within the vast expanse of the ocean. The waves surged, reaching almost to the very Throne of Glory. In response to this tumultuous display, the Almighty voice resonated through the heavens, commanding, “Be still, oh your waters!”

However, the waters, now calmed by the divine decree, became proud and boastful. They declared with arrogance, “We are the most potent force in all of creation. Let us overpower the Earth and claim it as our own!”

God, displeased by their arrogance, sternly admonished the waters, saying, “Do not revel in your strength, you vainglorious braggarts! I shall send forth sands to build a barrier against your ambitions!”

Upon witnessing the arrival of the sands, the waters, in their arrogance, shouted, “How can these tiny grains stand against us? A mere ripple of ours can easily engulf them.”

Yet, the grains of sand, though small and seemingly insignificant, were not disheartened. Their leader spoke words of reassurance to his brethren, saying, “Fear not, my fellow grains. It is true that each of us, in isolation, is but a tiny speck. The wind can carry us to the farthest reaches of the earth. But, if we remain steadfastly united, the waters shall bear witness to the true extent of our power!”

Motivated by their leader’s wisdom and words of courage, the tiny grains of sand came together from every corner of the Earth. They piled upon one another, forming formidable wall along the shores of the seas. These grains rose in mounds, built hills, and even formed majestic mountains, creating an immense barrier against the advancing waters.

When the waters beheld the formidable alliance of the united grains of sand, fear crept into their depths. Realizing the strength of unity, they retreated, acknowledging the indomitable power of even the tiniest of beings when they stand together as one.

And so, the secret of power lay not in individual strength, but in unity and solidarity, as demonstrated by the united grains of sand, who humbled the once-proud waters with their collective might.

The Foresighted Father

Once upon a time, in the Holy Land, there lived a wealthy merchant who had embarked on a lengthy ocean journey, accompanied by his faithful steward, a trusted slave. The merchant was a widower, and his only son remained behind, a pious young man deeply engrossed in the study of holy scriptures.

During the perilous voyage, fate took a dark turn as the merchant fell gravely ill. Sensing his impending demise, he decided to make his last wishes known. With a heavy heart, he drafted a will, bequeathing his entire fortune to his loyal steward. To his beloved son, he penned a message, offering him the choice of one single possession from his vast wealth.

After the merchant's passing, the steward-slave returned to the Holy Land with the will in hand. Facing the grieving son, he delivered the sombre news, saying, "Your father has departed from this world. He has left behind a will, designating me as the recipient of his entire fortune. However, he has graciously permitted you to choose any one item from among his many possessions."

The young son, heartbroken and confused, sought guidance from a wise rabbi in his community. With tears in his eyes, he recounted the story to the rabbi, detailing his father's final wishes. The rabbi, a man of great wisdom, listened intently and then spoke with a reassuring tone.

"Your father was a man of deep wisdom and foresight," the rabbi began. "He likely considered the possibility that if he left his entire fortune to you, the steward, in his role as guardian, might be tempted to take advantage. Instead, your father chose a different path. By naming the steward as the heir, he may have hoped that the steward, now the owner of his wealth, would protect it diligently, treating it as a precious treasure."

The rabbi's eyes twinkled with insight as he continued, "I advise you to follow your father's wisdom when you appear before the judges. When the steward presents the will, proclaim that your father granted you the privilege of selecting just one item from his possessions. Then, declare that you have chosen this very slave."

With a smile, the rabbi concluded, "Rest assured that if the slave belongs to you, all the wealth he has inherited from your father shall be yours as well, for the law dictates that whatever a slave owns ultimately belongs to his master."

Taking the rabbi's counsel to heart, the son followed his instructions when he faced the judges. The result was as the rabbi had foreseen: the young man gained possession of his father's entire estate, for the wisdom and foresight of his father's plan had ensured that his wealth remained within the family.

The Monkey-Mind Unleashed

Once there was a young man called Hari, who was a seeker after truth. He wanted to spend hours meditating and attain tranquillity of mind. He needed a Guru to teach him how to meditate without any distractions.

Hari set out to the mountainous Tibet. There was an old sage who lived in a temple on top of a hill. There he met his Guru. He was a saint who was known to have many mysteries, but it was very difficult to get any secret from him. He hardly ever spoke. The young man was determined to devote his whole life to his service and understand the mysteries. So, he decided to remain with the old sage.

After a few days, the sage told him, "You are unnecessarily wasting your time. Because I don't speak, people think I am keeping some secrets. I remain silent because I don't have anything to say. I don't hide any mysteries."

But the young man would not believe him. He said, "You will have to share with me the secret which opens the door of all the mysteries."

Finally, the old sage got fed up and wanted the young man to leave his presence. So, he told him, "Today I am going to tell you the secret. It is very simple."

The sage said the secret is a mantra which you must repeat a thousand times. He gave a mantra. "Keep repeating it, but there is a catch. Don't think of monkeys while saying the mantra."

Hari thought, "Oh, this is so easy! I have never thought about monkeys in my life. There is no reason why I should think about them now."

He rushed down the stairs of the temple where the old man lived. But strangely, even though he was not reciting the mantra, monkeys started coming, giggling. He would close his eyes and they would be there. While sitting, standing, eating, drinking. He even had nightmares of monkeys in his dreams! He could see only monkeys and monkeys. The more he pushed them away, the more they jumped towards him.

Within three days he was so bothered by the monkeys, he ran back to the saint on the fourth day. Hari asked him, "Take your mantra and the monkey back! 'I don't want them! Relieve me of this monkey business! I have gone crazy! I can't do anything!"

The old man said, "That's why I was not telling anybody because the key is very difficult. Now, do you understand why I was silent?"

The sage continued "Drop the mantra. Never ask for a formula for life, there is no formula."

Hari dropped the mantra, there was no monkey at all. He closed his eyes and there was no monkey. He looked all around and there was no monkey.

You cannot repress or suppress any thought. Never force any thought to go away; otherwise, it will rebound with greater energy.

A Circus of Unity

In a Bavarian town, during a gloomy period when Hitler's Nuremberg laws against Jews were in full force, a travelling circus arrived, its posters plastered all over the town square, praising a peculiar and dangerous attraction. The main event was to feature a man donned in the skin of a lion, who would boldly enter a tiger's cage to wrestle with the fierce predator.

Despite the circus's call for applicants to take on this perilous role, only one man showed up: a Jew, unmistakably marked by the degrading yellow badge on his arm.

The circus manager, clearly taken aback, exclaimed, "You're a Jew!"

The applicant, with a bitter resignation in his voice, responded, "Who else but a Jew would accept such a job? No one will employ me because of my heritage. I have no choice but to take this chance for the sake of my starving family."

And so, the Jew was reluctantly hired.

On the opening day, a massive crowd gathered, eager to witness the much-anticipated main attraction. The circus tent was filled, and excitement was in the air.

When the time came for the main event, the Jew made his entrance, trembling visibly. He adorned a lion's skin and, with a roar that echoed through the tent, he crawled on all fours and opened the cage of the fearsome tiger. Face to face with the menacing predator, his fear was palpable, and as he locked eyes with the tiger's cruel green gaze, he was overwhelmed with dread.

"It's all over for me now," he thought, and in the face of impending danger, he cried out in an unearthly voice, reciting the creed that Jews utter when confronted by death: "*Shema Yisrael!* Hear, O Israel!"

The tiger, to the astonishment of the lion, responded resolutely, "*Adonai Eloheinu Adonai Echad!*

The Lord our God, the Lord is one."

Startled and bewildered, the Jew turned to the "tiger" beside him, exclaiming, "You've frightened me out of my wits! I thought you were a real tiger!"

The "tiger" chuckled and retorted, "Listen, my friend. What makes you think you're the only Jew in Germany trying to make a living?"

In this tale, set against the backdrop of prejudice and discrimi nation, the circus performer finds an unexpected companionship with a fellow Jew, showing that even in the face of danger and adversity, unity and shared identity can bring strength and resilience.

The Bond of Kindred Spirits

Once upon a time, a young boy named Jim came across a signboard that read, "Puppies for Sale." His eyes sparkled with curiosity, and he couldn't resist the temptation to see them. Approaching the store owner, he asked politely, "May I see your puppies?"

In response, the store owner beckoned the mother dog and her litter of five adorable puppies towards Jim. As they scampered about, full of energy and life, Jim's attention was drawn to the last one in the group. This little puppy was different; he was limping, smaller than the others, and clearly had some physical challenges. However, Jim couldn't take his eyes off him.

"This is the one I want," Jim declared with determination.

The store owner hesitated; his concern was evident. He tried to dissuade the boy, saying, "Son, you might not want this puppy. He'll never be able to run and play with you like these other puppies can. He's had this disability since birth."

But Jim's heart was set on the little puppy, and he wouldn't be swayed.

Resigned, the store owner sighed, "Look, you don't have to buy that puppy. I'll give him to you if you really want him."

But Jim was resolute in his decision. "I don't want you to give him to me," he insisted. "I'll pay the same price as the other puppies."

Perplexed, the store owner inquired further, "But why do you want to buy a puppy with a disability?"

With determination, Jim rolled up one leg of his trousers, revealing a steel brace running down both sides of his leg, attached to a specially made shoe. Looking up at the man, he explained, "You see, sir, I don't run too well myself, and he'll need someone who understands."

Touched by the boy's empathy and understanding, the store owner picked up the little puppy and carefully handed him to Jim.

Jim inquired about the price, asking, "How much?"

With a warm smile, the store owner replied, "No charge. There is no charge for love."

But Jim insisted on leaving the money on the counter and walked out of the shop, cradling his new puppy in his arms.

The Clever Jackal and the Rats

Once upon a time, the king of the rats lived in a big hole at the edge of a jungle. Every day, the king and his subjects, hundreds of them, came out of the hole to look for food. Sometimes, they went deep into the jungle and at other times, to the nearby villages.

One day, when they were looking for food in the jungle, a jackal saw the rat king. He wanted to eat him. However, on second thought, he stopped himself from eating the rat king. He thought that rat king could not be alone. His subjects, ministers and family members must also be there in the forest. So, he decided to wait and see.

Soon, a huge pack of rats came and bowed before their king. They told him that they had gathered a large amount of food, and it was time to go home.

The jackal resisted himself from eating them. He thought, "I must find their dwelling place. Then I can eat a rat every day." So, he followed the rats to their hole at the edge of the forest. Once, he found their dwelling place, he planned to eat a rat every day.

Next day, when the rat king and his subjects were moving out of the hole, he stood up near the hole on his hind legs and looked upwards with his mouth wide open. The rat king saw him and asked him what he was doing there in such an uncomfortable position.

The jackal replied, "Already, mother earth is overburdened with the weight of so many creatures. Therefore, I thought of standing on my hind legs. I am looking upward to the heavens for the protection of this planet. I keep my mouth open so that I may drink in all the air I can. I have given up killing animals".

The rat king thought him to be a saint and ordered his subjects to offer him fruits and flowers. This was exactly what the jackal had wished for. When the rats came to offer their prayers, he asked them to come in a queue. Every day, as the rats were leaving, the jackal caught and ate the last one in the queue. Those who were in front did not know that the last one had been eaten. This went on for around a month.

The rat king noticed that the saintly jackal was becoming fat. He wondered how he could grow fat by feeding on air. At the same time, the rats noticed their number dwindling. Many of the rats were missing. The prime minister reported the matter to the king.

The rat king was very wise. He saw through the trick of the saintly jackal. Next day, he was determined to find out. After offering prayer to the jackal, he hid himself behind a bush and kept looking at the jackal. When all the rats had finished offering prayers, the jackal grabbed the last rat and ate him up. The mystery of the missing rats was solved.

The king called a meeting of the rats and told them the truth about the jackal. They all decided to attack him the next day. Next day, when they went to offer their prayers, the rat king jumped over the jackal's neck and bit him hard. The chief of rats bit him all over his body. Together, the rats killed the jackal and feasted on his body for several days. The wise rat king saved his entire clan from being wiped out.

And the rats lived happily in peace and quiet thereafter.

A Dream Come True

Aruna, a young woman of remarkable beauty with a flawless face, perfect teeth, and a figure to match, harboured a fervent ambition: she longed to become a superstar. In her high school days, she had dabbled in acting, participating in various skits and performances. As she grew older, she managed to secure some minor roles in a handful of films, but her dreams of stardom remained elusive.

One day, while strolling alone through a serene forest, Aruna chanced upon a crystal-clear pond. Enchanted by the tranquillity of the surroundings, she paused to gaze at her reflection in the shimmering water. A longing swept over her, and she whispered to herself, "Oh, how I wish I could be a star."

To her astonishment, a voice responded, "I can make you a star."

Startled, Aruna looked around but saw no one nearby. Confused, she inquired, "Who are you?"

Once more, the mysterious voice replied, this time emanating from the pond itself, "I can make you a star."

Intrigued and increasingly curious, Aruna peered down into the water, only to discover a smiling green frog gazing back at her.

"I'm a world-famous producer who has been transformed into a frog by a woman who I hired to play a witch, and who turned out to be real witch and she cast this spell on me," the frog said.

The actress, still in disbelief, questioned the frog, "You claim to be a world-famous producer who was transformed into a frog?"

The frog nodded enthusiastically. "Exactly! And if you trade places with me, the spell will be broken. I will regain my status as a world-famous film producer, and you, my dear, will have the opportunity to appear in all of my films."

Eager for her chance at stardom, Aruna agreed without much hesitation. "Okay," she said, her eagerness blinding her to any potential consequences.

With a magical incantation, the exchange of places was accomplished. The producer frog transformed back into a human, while Aruna found herself trapped in the form of a frog.

"But now when do I get back into human form too?" asked the actress frog.

Puzzled and dismayed, the newly restored producer shrugged and said, "Well, honey, that's the catch."

True to his word, the producer made a name for himself once more, creating numerous successful films. He was known for a peculiar quirk – there was always a role for a frog in his films.

And so, while Aruna had achieved her dream of stardom, it had come at an unexpected and rather amphibious cost. She became a permanent fixture in the producer's films, though in frog form, reminding her that dreams could sometimes take on the most unconventional shapes.

In the Orchard of Opportunities

Once upon a time, in a vast and mountainous landscape, an old female monkey embarked on a journey with her three beloved sons. Their travels led them from one towering mountain to another, fuelled by a sense of curiosity and adventure that burned within them. But alas, their wanderings eventually led them into unfamiliar terrain, and they found themselves hopelessly lost in the wilderness!

To compound their dire situation, the unforgiving winter had set in, and the once-abundant fruits, nuts, and nourishing sustenance had become a distant memory. The mountain upon which they now stood was barren and bald, offering no solace for their growling stomachs or their desperate need for warmth.

With the weight of her responsibility as a mother, the old monkey led her three sons to a secluded spot to kneel and pray. They begged the Mighty Lord for a glimmer of hope, for a chance to survive in this harsh and unforgiving landscape.

As they rose from their heartfelt prayers, a miracle unfolded before their eyes. A magnificent peach orchard materialized right in front of them, as if summoned by their deepest desires. Out from this orchard walked a benevolent gardener, extending an invitation to the famished monkeys.

He spoke kindly, "You may enter my orchard, dear monkeys, and each of you may choose a peach to eat. You are welcome to select the largest, ripest peaches you desire. However, remember this: each of you may only pick one peach. Furthermore, once you enter, there is no turning back or retreating. You must continue forward through the orchard, and when you reach the exit at the back, you must exit without hesitation."

The four hungry monkeys rejoiced at this unexpected bounty and eagerly entered the orchard, where they found peach trees laden with fleshy fruits. Abiding by the gardener's rules, they examined each peach meticulously, seeking the largest and most flawless ones to satisfy their hunger.

In due time, the mother monkey and her two older sons had chosen their preferred peaches, feasted upon them, and happily exited the orchard from its rear gate. However, the youngest monkey, though surrounded by big peaches, remained unsatisfied. He had diligently inspected each one and found flaws in all of them. Driven by the hope of discovering a perfect, flawless peach further ahead, he pressed on, leaving his family behind.

Alas, the perfect peach he sought remained elusive, and he continued his journey through the orchard until he spotted the exit sign. Unable to turn back, he reluctantly followed his mother and brothers and exited the orchard.

As they regrouped and shared their stories, the monkeys were astonished to discover that the peach orchard had vanished, as though it had never existed. Their adventure had taught them a valuable lesson: opportunities, once presented, may not come again. In their pursuit of perfection, they had lost the chance to savour the abundant blessings they had been granted, a lesson they would carry with them through life's unpredictable journey.

The Bread of Blessings

Once there was a young man, Joseph, who lived on a small farm. He was hard working, but poor. After his marriage he realized what he was earning was not enough to make both ends meet. There was not any job around for a man to do.

One day, Joseph said to his wife, Ann, "Honey, I will leave for a faraway place. I will get a job and work hard to be able to come back and give you the comfortable life that you deserve. I don't know how long I'll stay away. I won't be able to keep in touch with you. I only ask one thing, please wait for me. You should be faithful to me, as I will be faithful to you."

They agreed on that and, Joseph left. He walked many days until he found a very rich farmer, Paul, who needed someone to help him. Joseph offered his services. He was accepted.

Before starting his duties, he discussed his terms with his employer, "Sir, let me work for you as long as I want, and when I think I should go home, please relieve me of my duties. I ask you to save my salary for me, until I leave. The day I decide to go, please give me the money and I will go my way."

And they made a deal.

So, Joseph worked for twenty years without holidays and without any complaint. After twenty years he came to the farmer and said, "Sir, I want my money because I am going home."

Paul replied, "All right. I made a deal with you, and I will stick to it. However, before you go, I want to offer you something new, "I will give you all your money and send you away, or I will give you three pieces of advice and send you away. If I give you the money, I don't give you the three pieces of advice; and if I give you the three pieces of advice, I won't give you the money. Now go to your room and think about your answer."

Joseph thought hard for two days. Then he went to Paul and told him, “I want the three pieces of advice.”

Paul stressed again, “If I give you three pieces of advice, I won’t give you the money.

And the man replied, “I want the three pieces of advice.”

The boss then told him,

1. “Never take short cuts in your life; shorter and unknown paths could cost you your life.
2. Never be too curious, for curiosity towards evil can be deadly.
3. Never make decisions in moments of anger or pain, because when you repent it could be too late.”

After giving these three pieces of advice Paul said to him, “Here you have three loaves of bread; two are for you to eat during the journey and the last is for you to eat with your wife when you get home.”

So, the man went on his way, after twenty years away from home and from his wife, whom he loved so much. After the first day of travel, he found a man who greeted him and asked, “Where are you going?”

He replied, “To a distant place which is about 20 days away if I continue walking on this road.”

The man said to him, “Oh boy, this path is too long! I know a short cut that is very safe, and you will arrive in five days.”

The man began to follow the path suggested until he remembered the first piece of advice.

“Never takes short cuts in your life; shorter and unknown paths could cost you your life.”

Then he returned and followed the long path. Days later, he learned that the shortcut led to an ambush for travellers, where they would be deprived of all their belongings and even their very lives.

After a few days of travel, he found an inn by the roadside where he could rest. He paid for the room, and, after taking a bath he lay down to sleep. During the night he woke up as he heard a terrifying scream. He rose to his feet and went to the door to check what had happened. When he was opening the door, he remembered the second piece of advice, "Never be too curious, for curiosity towards evil can be deadly." Therefore, he returned, lay down and slept.

At dawn, after having his breakfast, the owner of the lodging asked Joseph if he had not heard the scream at night. He affirmed that he heard.

Then the host said, "Were you not curious to see what happened."

And he replied, "I was not."

Now the host said, "You are the first guest to leave this inn alive. My neighbour is completely crazy. He usually shouts at night to get someone's attention. When some of the guests come out, he kills them and buries their body in the woods close by.

Joseph continued his long journey eager to arrive soon. After many days and nights walking, he was very tired, but he finally saw his house at a distance. It was night. He saw some light coming out from the window of his house and was able to see the silhouette of his wife, Ann. But he was also able to see that she was not alone. He came closer and saw there was a man with her. She softly caressed his hair. When he saw that scene, his heart was filled with hatred and bitterness. He decided to rush against the two and kill them mercilessly. However, he took a deep breath, and he remembered the third piece of advice, "Never make decisions in moments of anger or pain because when you repent it could be too late."

Then Joseph stopped, reflected, and decided to sleep outside that night. He slept amid the bushes and decided the course of action on the next day. At dawn, he was calmer and said, "I will not kill Ann and her lover. I am going back to my boss and ask him to take me back. But before that I want to tell Ann that I have always been faithful to her."

Joseph went to the front door and knocked. When Ann opened the door and recognized him, she cried and embraced him warmly. He tried to push her away, but he was not able. Then with tears in his eyes he told her, "I was faithful to you, but you betrayed me."

The accusation shocked Ann. So, she replied, "How did I betray you? I have never betrayed you. I waited patiently for you for twenty years."

Then he asked, "How about the man that you were caressing yesterday?"

And she said, "That man is our son. When you left, I discovered I was pregnant. Today he is twenty years old." Hearing that Joseph asked her forgiveness. He ran in and hugged his son. After that he shared with her all his experiences when he was away.

Meanwhile, Ann prepared some coffee for them. Joseph took out the packet of bread given by his employer and placed it on the table saying, "This is all I have earned in twenty years." After a prayer of thanksgiving, he broke the bread. When he looked at it, he found many gold coins inside. In fact, there was even more than the right payment for his twenty years of dedication and hard work.

Unequal Blessings

In a faraway land, there lived a humble donkey, Levi that toiled tirelessly under a poor washerman. Day in and day out, it bore the heavy loads of laundry, trudging to the riverbank for washing. Levi's existence was marked by meagre sustenance, and it was a mere shadow of the robust animal it could have been.

One day, fortune smiled upon the weary animal. The chief of the Sultan's stables, a man of compassion, came across the pitiable creature. His heart went out to the poor Levi, and he resolved to rescue him from his wretched life. With kindness in his heart, he ordered the donkey to be brought into the royal stables, where it would be well-fed and cared for.

Levi, now amidst the majestic Arab steeds, marvelled at the stark contrast between his past and present existence. The royal horses were a sight to behold, their glossy coats and strong bodies radiating health and vitality. Levi, however, had seen neither good food nor gentle care for a long time.

In its moments of contemplation, Levi turned to the heavens and questioned why such an inequality existed, why its life had been so wretched while these regal horses lived in opulence and splendour.

As days went by, Levi gradually got used to his new life. Now he was well fed. People were assigned to take good care of him, and he slowly started to regain his health. But he was always complaining to God for his unfairness, since he always felt inferior among the robust and colossal war horses.

Meanwhile tensions over the border dispute were on the rise between the Sultan and the ruler of the neighbouring country. And not long afterwards, war seemed inevitable, casting a shadow over its once-peaceful corners. The Sultan's prized Arab horses, symbols of grace and strength, were summoned for the battlefield. They left the stables, their noble heads held high, and did not return for many months. Levi was wondering what had happened to those horses. After a while, when the war came to an end, some of the horses returned to the stable.

In the cruel aftermath of battle, the Arab horses returned, bearing the scars of war. Their graceful bodies were now marred by the painful evidence of conflict – arrows embedded in their flesh; wounds covered in bandages. Some, tragically, did not return at all, having given their lives in service to the Sultan.

The sight of the wounded and the thought of the fallen Arab horses, once so magnificent and unblemished, served as the poignant and profound response to the disparity he noticed around him. Levi was deeply troubled by the suffering he saw around him. So, he looked up once again, but this time to thank God for keeping him away from such torments and agony. And this was indeed the answer to his question to God about the unfairness he had felt.

He understood now that life's inequalities and the stark contrasts between beings could change in an instant. Fate was fickle, and what appeared as misery might, in the blink of an eye, be transformed into blessings. Levi had witnessed the impermanence of circumstance and the transient nature of fortune, and he learned that gratitude for the present moment was the key to finding contentment in life.

The Monk and the Wolf

In a distant forest, a hunter roamed the woodlands in search of game. The morning had passed, and he had yet to have any success with his bow. Frustration gnawed at him until, finally, his eyes fell upon a lone wolf. Determined to make a catch, he drew his bow and aimed true, releasing the arrow that found its mark. The arrow struck the wolf, but it managed to flee, vanishing into the dense woods.

Not far from this scene, an aged monk moved slowly along a woodland path, bearing a heavy bag filled with crops. The wounded wolf, consumed by fear and agony, suddenly emerged before the monk. The wolf, its pain evident, humbly implored the monk for aid.

The wolf knelt before the monk and pleaded, "Kind Sir, you have always shown mercy to the creatures of this forest. I am grievously wounded and in desperate need of your help. A heartless hunter has injured me, and he is hot on my trail. If you spare my life, I will forever be in your debt and repay your kindness in any way I can."

Without hesitation, the compassionate monk emptied his bag, allowing the wounded wolf to seek refuge within it. In a matter of moments, the pursuing hunter arrived at the scene. He inquired of the monk whether he had seen a wounded wolf, but the monk replied, "No," and the hunter continued his way.

Once the hunter had departed, the monk gently opened his bag and helped the wolf out, removing the arrow that had pierced its back.

The wolf, now free and recovered, gazed around at the tranquil forest, where only the monk remained by its side. However, instead of gratitude, the wolf's eyes filled with a chilling hunger. It turned to the monk and declared, “Thank you for your aid, but now I must ask for more. I have not eaten in days, and my hunger is insatiable. I must consume you, or I shall perish from starvation.”

Shocked and disheartened, the monk protested, “How can you repay my kindness with such cruelty?”

The wolf, unmoved by the monk’s pleas, insisted, “You should not have saved me. It is too late now. I am a wolf, injured and famished. There is no other sustenance for me here. I must devour you.”

As the wolf prepared to attack the monk, the old man suggested, “Wait a moment. Let us find a third party to arbitrate our dispute. If this impartial judge agrees with your claim, I shall accept my fate willingly.”

The wolf reluctantly agreed, convinced that the monk would not escape. Together, they ventured back to the main road, where they soon encountered a shrewd politician. The monk recounted the entire story, seeking a fair judgement.

The crafty wolf seized the opportunity and accused, “When the monk placed me in his bag, he intentionally sealed it so tightly that I nearly suffocated. He intended to murder me.”

The monk swiftly defended himself, explaining that he had to close the bag securely to deceive the pursuing hunter.

Both parties attempted to sway the politician with their arguments, leaving him torn between two conflicting tales.

“As a politician, I trust only what I can see,” declared the politician. “I cannot render judgement unless the wolf re-enters the bag, allowing the monk to close it once more. Only then will I assess whether the monk acted wrongly.”

Both the wolf and the monk consented to this proposal. The wolf obediently entered the bag once more, and the monk securely fastened the opening, just as before. The politician took further precautions, tightly binding the bag’s mouth with a sturdy rope to ensure the wolf remained inside.

The politician, now in a position to adjudicate, proclaimed, “In my role as judge, I find the wolf guilty of malevolent intent and action. He shall receive the penalty of death for his wicked thoughts and deeds."

Before the wolf could utter a protest, the politician-judge promptly drew his knife and carried out the sentence on the spot, putting an end to the matter.

Instinct and Compassion

In the calm hours of early morning, a wise hermit followed his daily ritual by descending to the river for a bath. As he stood by the riverside, he noticed a scorpion struggling in the water near him. The sage knew that scorpions cannot swim and if he did not save the scorpion, it would drown.

Moved by compassion, the saint reached out to rescue the troubled creature. Carefully, he picked up the scorpion, saving it from the impending doom of drowning. But as he gently placed the scorpion on the riverbank, it responded to its instincts and stung the saint's finger. A sharp, searing pain shot through the sage's hand, causing him to involuntarily recoil, accidentally sending the scorpion back into the water.

After regaining his composure, the saint, determined to help the scorpion, reached for it once more, only to be met with the same result – another painful sting. This pattern repeated itself several times as the sage persisted in his efforts to save the scorpion, each attempt met with a venomous sting.

The sage's disciples, who had been observing the unfolding drama, became increasingly concerned for their beloved teacher. They begged him to abandon the perilous attempt and save himself from the relentless attacks of the scorpion.

But the sage paid no heed to their pleas and continued his mission to rescue the creature, undeterred by the pain it inflicted upon him. He persevered until he finally succeeded in securing the scorpion's safety on the riverbank.

The disciples, carefully tended to the sage's wounds and waited for him to recover. Once he regained consciousness, one of his devoted disciples inquired, "Revered master, we are deeply perplexed. Why did you allow the scorpion to sting you repeatedly while you were trying to save it?"

With a serene smile, the sage replied, "My dear child, the scorpion did not sting me out of malice or ill will. Just as it is the nature of water to make things wet and fire's nature to burn, it is the nature of a scorpion to sting. The creature, in its ignorance, continued to sting me because it did not understand that I was attempting to rescue it. But just as it is the scorpion's instinct to sting, it is my nature to be compassionate and to save. In the same way the scorpion could not forsake its nature, why should I abandon mine?"

The Frogs' New King

Long, long ago the frogs were living a very free and comfortable life among the reeds beside lakes and ponds. They were discontented as they had no king to rule over them. They prayed to Jupiter the supreme god of gods for a king to rule over them. Jupiter was having a party along with other gods and he was annoyed by the frogs' croaking. To regain his peace, he dropped a huge log into their pond. The splash terrified them.

The frogs watched their king in great awe. They hid themselves among the reeds and grasses, thinking the new king to be some fearful giant. But they soon discovered how tame and peaceable king Log was. In a short time, the younger frogs were using him for a diving platform, while the older frogs made him a meeting place, where they croaked loudly to Jupiter about the government.

Jupiter wanted to teach the discontented frogs a lesson. The next day, the frogs awoke to see their magnificent king in the pond. It was a white bird, a crane, with long, slender legs and an impressive beak. It looked elegant and beautiful indeed.

The frogs rejoiced at their new king. Finally, there was one they could be proud of. Not only was he beautiful, but he was also fearsome with his long beak and sharp claws. The frogs were beside themselves with joy. They jumped up and down.

Pretty soon, the frogs got bolder. They started to climb up their king's long legs and up his feathery winged body. Some even dared climb that long, elegant neck. This was a triumphant moment indeed. The frogs croaked away noisily but happily.

Suddenly, the crane reached into the pond and grabbed a fish. Next, it reached again into the pond, catching a frog this time, and swallowed it whole.

This was not how the frogs imagined him to be. The king was supposed to protect them, not eat them! At this point, the frogs, and everyone else in the pond, realized that if they did not scramble out of the way, they would be next on the menu. All the joyous noise earlier was now replaced by frantic croaking. The crane continued to feast on everything in sight and within reach. The day of rejoicing had turned into a day of sorrow for the frogs.

The next day, the pond was quiet and empty. The frogs had either been eaten up by the crane or escaped into a nearby pond. Those frogs never did ask Jupiter again for another king.

So, if you ever hear frogs croaking non-stop, don't be surprised. They are just telling their story to anyone who cares to listen. Their story also comes with a warning: Be careful what you wish for: You might just get it!

Learning the Craft

The son of a master thief asked his father, to teach him the secrets of the craft. The master agreed, and that night took his son to break into a large house. While the family was asleep, he silently led his young apprentice into a room that contained a clothes cupboard. The father told his son to go into the cupboard to pick out some clothes. When he did, his father quickly shut the door and locked him in.

Locked inside the chest, the boy was flooded with emotions - anger, fear, and confusion swirled within him. He pondered how to escape from his seemingly dire situation. After a while, a spark of inspiration ignited within him. He began to mimic the sound of a cat, emitting soft, desperate meows.

The homeowner, awakened by the noise, fetched a candle, and cautiously approached the locked cupboard. As he opened it, the boy seized the opportunity and slipped out, swiftly extinguishing the candle, and without wasting a moment, he made a hasty escape.

Hours later, the exhausted son returned home, his feelings a mix of anger and relief. He confronted his father, demanding to know why he had locked him in that cupboard. "Father," he cried out, "Why did you trap me like that? If it weren't for my desperation to escape, I might never have found a way out. It pushed me to think creatively and quickly!"

The old thief smiled proudly at his son and said, "My dear, you've just experienced a crucial lesson in our craft."

"A challenge," he continued, "brings out the creativity in a person. Your mind operates at its best and quickest when you're pushed to your limits, with your back against the wall."

This story emphasizes the importance of resilience, quick thinking, and the ability to adapt when faced with adversity, rather than glorifying burglary.

The Crutches

Once upon a time, in a quaint island village, there lived a wise and respected village chief. However, his life took an unexpected turn when he met with a grave accident that left him to walk on crutches. Despite the misfortune, the chief's indomitable spirit refused to wane.

As time passed, a peculiar thing happened. The chief, with his crutches as his constant companions, began to develop remarkable agility. He transformed his limitations into opportunities and became an expert at moving with astonishing speed. Moreover, he started learning to dance, perform tricks, and entertain his friends and neighbours. The villagers marvelled at the chief's skill on crutches.

The chief did not stop at just mastering the art of crutch-walking; he imparted his knowledge to his children. Soon, the entire village was buzzing with envy as they watched the chief and his children navigate effortlessly with their crutches. Eager to imitate their leader, the villagers began to practise crutch-walking diligently. In time, all the villagers became proficient in the use of crutches, and talented craftsmen emerged to craft exquisite crutches from materials like gold, silver, and ivory for the affluent members of the community.

Years passed, and a peculiar transformation overtook the village. Gradually, nobody in the village could be seen walking without crutches.Even visitors who came from neighbouring villages adopted the practice. It became a way of life, and no one questioned it.

One day, a young man arrived at the village. Unlike the villagers, he had never seen people walk exclusively on crutches. He could not help but be curious. As he observed the villagers, a growing doubt troubled him. "Why do they all use crutches when they have perfectly good legs?" he wondered.

Determined to unravel this mystery, the young man decided to confront the village elders. With unwavering confidence, he demanded an explanation for the peculiar custom. The elders, amused by this audacious newcomer, decided to put him to the test.

"Why don't you demonstrate it to us if you think you know better?" they challenged him.

Without hesitation, the young man agreed. A date was set for a grand demonstration in the village square, and excitement rippled through the community. On the appointed day, the entire village gathered, eager to witness this bold newcomer's attempt to defy their long-standing tradition.

As the village clock chimed, signalling the moment of truth, the young man hobbled to the centre of the square using his crutches. With a dramatic flourish, he stood upright and, to the collective gasp of the crowd, dropped his crutches. Silence enveloped the square as he took a brave step forward.

But alas, as if the very laws of nature protested his defiance, he stumbled and fell flat on his face.

The villagers watched in stunned disbelief, their faith in crutches reaffirmed. The young man's attempt had only strengthened their conviction that walking without crutches was an impossible feat.

And so, the village continued to live its peculiar reality, where crutches were not just tools of support but symbols of their unquestioning faith in tradition.

Bounty Beneath the Oven

In the bustling city of Krakow, there lived a poor Jew named Isaac. Night after night, he was haunted by a recurring dream that whispered of a hidden treasure concealed beneath the grand royal bridge in faraway Prague. At first, he dismissed these visions as mere fantasies, believing they were beyond his reach. Yet, the dreams persisted, as vivid and insistent as the stars in the night sky. To Isaac, it felt like a divine calling, a message from above guiding him towards an unimaginable destiny.

With unwavering determination, Isaac decided to heed this mysterious call. He embarked on a perilous journey, leaving behind his humble life in Krakow, and setting out towards the distant city of Prague. Upon his arrival, he discovered that the bridge he sought was heavily guarded by vigilant soldiers, their watchful eyes always tracing every part of the bridge. He approached the bridge from different angles, seeking an opportunity to unearth the hidden treasure, but the ever-present scrutiny made it an impossible task.

His quest seemed fruitless until he caught the attention of an officer patrolling the area. Suspicion clouded the officer's gaze as he observed Isaac's constant presence near the bridge. Feeling cornered and out of options, Isaac decided to confide in the officer, sharing his haunting dreams and the purpose of his journey. He hoped against hope that the officer might offer guidance or assistance in realizing his divine vision.

However, the officer's reaction was far from what Isaac expected. He burst into laughter, his amusement echoing through the streets of Prague. Once he had composed himself, he turned to Isaac and said, "Look here, my friend, I, too, had a dream. In my dream, a man named Isaac from Krakow possessed a treasure hidden beneath his very own oven. Do you see me rushing to Krakow to unearth it? Of course not! Go back to your home and let go of these dreams; they are nothing more than fantasies!"

Isaac listened intently to the officer's words, his heart pounding with a mix of confusion and hope. He realized that perhaps the officer's dream held the answer he had been seeking all along. With newfound determination, he wasted no time in retracing his steps back to Krakow. When he reached his modest abode, he knew precisely what he needed to do.

He dug beneath his oven, his hands trembling with anticipation. To his astonishment and overwhelming joy, he unearthed a hidden treasure that had been waiting patiently beneath the very heart of his home. The riches he discovered were beyond his wildest dreams, a testament to the power of his unwavering faith and the mysterious ways of the world.

As the legend goes, the wealth Isaac uncovered in his own humble dwelling was used to build the magnificent Isaac Synagogue in Krakow, a symbol of enduring faith and the realization that sometimes, the greatest treasures lie within us and the places we call home. It serves as a reminder that the most profound journeys may lead us not to distant lands, but to the depths of our own souls.

The Mousetrap

Long ago, a curious mouse happened to peer through a small hole in the wall, where he watched the farmer and his wife unwrapping a mysterious package. The mouse's inquisitive nature led him to wonder about the contents of this parcel. However, his anticipation quickly turned to dread when he realized that it contained a mousetrap.

The mouse was seized by panic and darted up and down, shouting out a warning to all the creatures in the house, "There is a mousetrap in the house! There is a mousetrap in the house!"

Hearing the mouse's frantic cries, the chicken, who was busy clucking and scratching for grains, paused and raised her head. She said in a very casual tone, "Mr. Mouse, I understand your concern, but this matter doesn't affect me in the least."

Disheartened by the chicken's indifference, the mouse hurried over to the pig and urgently repeated his warning, "There is a mousetrap in the house! There is a mousetrap in the house!"

The pig, displaying sympathy for the mouse, replied, "I'm truly sorry to hear that, Mr. Mouse, but there is nothing I can do to help you except offer my prayers."

Undeterred, the mouse turned to the cow and conveyed the same urgent message, "There is a mousetrap in the house! There is a mousetrap in the house!"

The cow, while expressing her condolences, said, "I'm sorry for your predicament, Mr. Mouse, but this situation doesn't concern me."

With heavy heart, the mouse returned to the house, his hope of assistance from his fellow creatures shattered. He braced himself to face the farmer's menacing mousetrap all alone.

That very night, the farmer and his wife set the mousetrap, eager to capture any unwelcome intruder. The following morning, the farmer's wife made her way to the trap to inspect their catch. In her haste, she failed to notice that the trap had ensnared a venomous snake by its tail. As she approached, the snake, acting in self-defence, bit the farmer's wife. Her condition rapidly deteriorated, and the farmer rushed her to the hospital. She returned home with a fever.

Days passed, and the farmer's wife's health showed no improvement. The farmer wanted to treat his ailing wife with fresh chicken soup. The farmer took his hatchet to the farmyard for the soup's main ingredient. But his wife's sickness continued. So, friends and neighbours came to sit with her round the clock support. To feed them, the farmer butchered the pig. Despite all their efforts, the farmer's wife's condition continued to worsen, and in the end, her illness proved overwhelming, and she did not recover, leaving the farmer devastated.

The news of her passing spread, and an even larger gathering assembled for her funeral. The farmer, determined to provide for everyone in attendance, slaughtered the cow to ensure there was enough meat to serve them all.

From his vantage point in the wall, the mouse observed these events unfold with a profound sense of sorrow. He realized that his warning about the mousetrap had not been heeded and the consequences had far-reaching effects that touched the lives of everyone on the farm.

The Mouse kept thinking,"When one of us is in trouble, it's everyone's problem. What happens to one can affect all of us, so we should all care and help when someone needs it."

The Power of Trust

Once a man named Hassan was travelling through the desert in a caravan. He had a purse full of coins. He wanted to entrust them to someone because he had heard that robbers were about.

He looked around and some way off, saw a tent, and a man was sitting there, a most distinguished-looking man. So, he said, "Will you please keep this purse? For I am afraid that if the robbers come, they will take it."

The man said, "Give it to me, I will keep it."

When Hassan came back to the caravan, he found that robbers had come and taken all the money of his fellow travellers, and he thanked God that he had given his purse to someone to keep.

But when he returned to the tent, he saw all the robbers sitting there, and among them was this most dignified man dividing the spoils. He realized that this was the chief of the robbers and thought, "I was more foolish than all the others, for I gave the money to a thief. Who can be more foolish than that!" And he was frightened and backed away.

But as soon as the chief saw him he called him and said, "Why are you going? Why did you come here?"

He said, "I came here to get my purse back, but I found that I had given it to the very band from which I wanted to protect it."

The chief said, "You gave me your purse, is it not so? You trusted it in me, and it was not stolen from you. Did you not trust me? How can you expect me to take it from you? Here is your purse, take it."

This act of trustworthiness impressed the robbers so much that they followed the example of their chief. They gave up robbery. It moved them to the depths of their hearts to feel what trust means. And in his later days, this chief accomplished great spiritual work.

The Whip that Saved a Life

Once upon a time, there lived a man named Ali, who found himself in an unusual predicament. As the sun dipped below the horizon, Ali, lost in the realm of dreams, unwittingly swallowed a scorpion. The venomous creature, finding itself trapped in the confines of Ali's throat, caused him to awaken in sheer panic.

Ali, in a state of delirium, sprang to his feet, coughing and convulsing to rid himself of the unexpected intruder. His desperate struggle attracted the attention of a passing horseman named Malik. Malik, a man known for his quick thinking and practical nature, assessed the situation in an instant.

Without a word, Malik raised his whip high and began to mercilessly lash out at the bewildered Ali. Whiplash after punishing whiplash rained down upon him, leaving Ali bewildered and in agony. In his delirium, Ali attempted to cry out for mercy, but his words were drowned by the relentless onslaught.

Malik, seemingly heartless, continued the assault, not uttering a single word. He knew that time was of the essence, and the scorpion's venom was a deadly threat that needed to be eliminated swiftly.

In his frantic struggle, Ali writhed on the ground, rolled over and over, but the whipping continued without pause. It was a bewildering and terrifying experience for Ali, who couldn't comprehend why he was being subjected to such cruelty in his time of need.

Finally, with a herculean effort, Ali's stomach rebelled against the venomous invader. With a violent heave, the scorpion was expelled and tumbled to the ground. It scurried away, leaving Ali in a state of exhaustion and relief.

As Ali gasped for breath and tried to regain his composure, he realized the truth. Malik's seemingly heartless actions had been the only way to ensure the scorpion's expulsion before its venom could claim his life.

Malik, without a single word of explanation or apology, spurred his horse and rode away into the distance, leaving Ali to contemplate the strange turn of events. In time, Ali came to understand that what had initially appeared to be an unjustified assault was, in fact, a lifesaving intervention that he would forever be grateful for.

Cooking up Melodies

In a kingdom known for its love of music, there reigned a king who had an insatiable passion for orchestral performances. To him, the grandeur of music lay in its volume, and he believed that the louder the melodies, the more splendid they were.

One fateful evening, during a grand banquet attended by nobles and dignitaries, disaster struck. One of the musicians, poised to perform in front of the king, injured his hand and was rendered unable to play. Panic ensued among the orchestra members, and the conductor, fearing the king's displeasure, raced backstage in search of a replacement.

As he anxiously paced behind the curtains, the conductor's ears caught an unexpected sound emanating from the nearby kitchen. It was a rhythmic and staccato noise unlike anything he had heard before. Intrigued, he rushed into the kitchen and beheld a cook swiftly chopping vegetables with impeccable precision and then slicing beef at a deliberate pace.

In a moment of inspiration, the conductor approached the cook with a proposal. He explained the dire situation and implored the cook to take the musician's place for the night's performance.

"But, Sir, I am but a humble cook," protested the bewildered cook. "Cooking runs in my blood—my father was a cook, and his father before him. Music, on the other hand, belongs to your world, not mine."

The conductor, undeterred, reassured the cook. "Mr. Cook, you need not worry. The king has no discerning ear for music; he is solely enamoured by his concubine dancers. He cannot differentiate between a master musician and an impostor. All you need to do is follow my instructions. When you hear fast-paced music, move your hands as if chopping vegetables swiftly. And when you hear slower notes, mimic the action of cutting tough beef. Just ensure you make no loud noises, and you will pass as a fine musician."

With reluctance, the cook agreed to the conductor's plan. That night, he donned the musician's attire and joined the orchestra. As the music played, he followed the conductor's cues, mimicking the actions of chopping vegetables and slicing beef, all while making no unnecessary sounds.

Over the years, the cook continued to participate in the orchestra, pretending to be a musician even though he had never truly acquired the skills. He became a regular performer, his trick undetected by the king.

Time passed, and eventually, the old king passed away, succeeded by his son, a music enthusiast in his own right. However, the new king had a different taste in music. He enjoyed listening to individual performers and relished the nuances of their artistry.

When it was time for each musician to perform for the new king, the cook, fearing his deception would be revealed, vanished overnight, leaving a legacy of culinary artistry that had unexpectedly delighted the royal orchestra.

Tasting the Apple

Once upon a time, in a classroom filled with eager students, a teacher decided to teach them a valuable lesson about apples. He placed a couple of fresh, enticing apples on his desk and told the class, "Today, we're going to explore the essence of apples. Whoever can describe them in the most meaningful way will have the opportunity to accompany me to the town."

The students were instantly fascinated and excited. They understood that this was not their typical lesson. Everyone attempted to describe the apples, eager to prove his knowledge.

The first student began by sharing the historical origins of apples, mentioning that they were believed to have been cultivated in Europe thousands of years ago. He delved into the fascinating journey of this beloved fruit through time.

The second student spoke passionately about the introduction of apples into India during the nineteenth century by the British. He described the impact of this introduction on the culture and cuisine of the region.

Another student highlighted the various practical uses of apples, such as their transformation into cider, their role in delicious desserts, and their use in making applesauce. The classroom buzzed with information and insights.

One by one, the students contributed their unique perspectives on apples, each shedding light on different aspects of this beloved fruit.

However, the last student took a different approach. Instead of words, he chose action. With a gleam of inspiration in his eyes, he took a knife, deftly sliced a piece of apple, and gently placed it into the teacher's mouth. With a subtle motion, he encouraged the teacher to chew, allowing the juicy apple to burst with flavour.

The teacher, tasting the crisp and sweet apple, nodded in approval. With his mouth still full, he managed to say, "Exactly. Apples cannot be fully described or explained with words alone. They must be tasted, experienced, and savoured. The true essence of apples is not found in textbooks or lectures; it is discovered when you take a bite and let their deliciousness speak for itself."

The students were filled with a newfound appreciation for apples. They shared the remaining apples among themselves, savouring every bite. And as a reward for his unique approach, the last student earned the chance to accompany the teacher on a trip to the town, knowing that some lessons are best learned through experience.

Clean the Window First

In a serene neighbourhood, a young couple had recently settled into their new home. One sunny morning after moving in, as they sat down for breakfast, the young woman glanced out the window and noticed her neighbour hanging her laundry to dry.

"Look at her laundry," the young woman remarked, a hint of disapproval in her voice. "It doesn't seem very clean. She must not know the right way to wash clothes. Perhaps she should try a better detergent."

Her husband, a man of few words, observed silently. However, every time their neighbour put her laundry out to dry, the young woman couldn't help but make similar comments about its cleanliness.

Several weeks passed, and one day, the young woman peered out the window to find a delightful sight: her neighbour's laundry was immaculate, clean, and bright. She turned to her husband, astonishment in her voice.

"Would you look at that?" she exclaimed. "It appears our neighbour has finally learned how to wash her clothes properly. I wonder who taught her."

With a sly smile, her husband responded, "Well, my dear, this morning I decided to get up early and clean our windows."

The Lost Goat Found

One fateful day, farmer Ravi made a shocking discovery. Several of his goats had disappeared without a trace. His initial anger turned to confusion, and he could not understand how this could have happened. As he pondered the mystery, he could not ignore the unsettling feeling that his two sons, Kiran, and Rahul might be involved in the goats' disappearance.

Summoning his sons to the front yard, Ravi's stern face was illuminated by the burning afternoon sun. In his hand, he held a long, menacing whip, a symbol of both his authority and his frustration.

"How many goats have you lost?" bellowed Ravi, his voice echoing through the countryside.

The elder son, Kiran, lowered his head and mumbled, "One, Father."

"And where were you when the goat ran away?" the farmer demanded, his eyes drilling into Kiran's soul.

With a heavy heart, Kiran confessed, "I was gambling with the other village boys, Father."

"Gambling!" roared the father, his patience exhausted as he lashed the whip onto Kiran's back.

Turning to his younger son, Rahul, Ravi asked, "And how many goats have you lost?"

"Just one, Father," Rahul replied, his voice trembling slightly.

"What excuse do you have for losing just one goat?" the farmer inquired, a mix of anger and disappointment in his eyes.

"I was studying, Father," Rahul replied with an air of righteousness.

"So what?" Ravi retorted sharply, swinging the whip once again, this time striking Rahul's back. "Even though your cause may have been nobler, it is no excuse for neglecting your duty to watch over our precious goats."

As the sun began to set, the lessons of that day were etched into the hearts and minds of both sons. They had learned that their actions, whether driven by neglect or indulgence, could have consequences for their family and livelihood. From that day forward, they vowed to be responsible towards their father's farm.

The goats, though fewer in number, continued to graze the lush fields, and the farm flourished once again. The family's bond grew stronger as they worked together to protect their herd, ensuring that no goats would be lost to neglect or folly ever again.

The Cubing Brilliance

Erno Rubik is a clever inventor, architect, and professor from Hungary. Early in his life, he was not widely known. Little did he imagine that his life was about to take a fascinating turn - literally! He was on the path to creating something truly iconic in history, the incredible Rubik's Cube!

The story begins in 1974 when Erno had a very clever idea. He made a cube not to be famous or rich but to help his students understand three-dimensional geometry better. This ordinary cube, though, turned out to be extraordinary. He didn't think his invention would become so famous worldwide with its simple yet tricky design. What started as a tool for teaching turned into a worldwide sensation - a symbol of smart thinking that amazed people everywhere.

The first time Erno decided to tackle the cube himself, it turned into a month-long twisting and turning marathon to align all the colours perfectly. Little did he imagine that his creation would soon become a worldwide phenomenon, capturing the hearts of puzzle enthusiasts everywhere.

As the Rubik's Cube gained popularity, a new challenge emerged. Enthusiasts worldwide engaged in a race against time, competing to solve the puzzle in the blink of an eye. Some possessed an almost magical ability to crack the cube in mere seconds, showcasing a blend of skill, memory, and reflexes. These marvels, known as "speedcubers," even went to the extent of memorizing complex algorithms to navigate the cube's intricate arrangements with lightning speed.

Despite being the mastermind behind the cube, Erno found himself in awe of these speedcubers. His creation had taken on a life of its own, surpassing his own abilities. The puzzle that once took him a month to solve was now conquered in seconds by others.

This historical tale is not just about a colourful cube; it is about the extraordinary potential within each of us to achieve the seemingly impossible. The Rubik's Cube became more than a puzzle; it is a symbol of creativity and trying new things. Erno Rubik's cube is a testament to how our minds are amazing, and we can do things we thought were very hard. And so, the Rubik's Cube continues to twist and turn its way through history a reminder that within simplicity lies the potential for boundless complexity.

The Jealous Goat

In a peaceful village, there lived a goat and a donkey on the same farm. The farmer relied on the donkey to help with various tasks around the farm, such as carrying loads and ploughing the fields. The donkey worked diligently and was always rewarded with a little extra food. The goat became jealous, forgetting all the donkey's hard work. He thought if the donkey stopped working, he would get his food.

The goat's envy ran wild. He wanted to play a trick on the poor beast of burden. The goat said, "You slog the entire day. But our master is very cruel to you. He beats you every day and places heavy burdens on your back and you stagger underneath. Don't you want some rest?"

"I do," replied the donkey enthusiastically.

"Then do as I say. Pretend you are sick and fall unconscious. The master will look after you well and you will get rest for a few days."

The donkey liked the idea. He followed the advice of the goat. He lay down as though he was very sick. The farmer called the village vet.

The vet examined the donkey and scratched his head, puzzled by the unusual illness. He finally suggested, "I have a remedy for this ailment. Feed the donkey a special soup made from goat meat, and he will recover."

The farmer, trusting the vet's advice, decided to slaughter the goat to prepare the soup. The goat understood his own death was the result of his jealousy. Before he had the time to bleat, the farmer cut his sharp knife into the goat's throat. The goat finished up as food for the donkey!

The Unlikely Companions

In a distant forest, there lived a majestic lion. One scorching afternoon, as he went to a river to quench his thirst, an unfortunate mishap occurred. His powerful paw became ensnared in the muck on the riverbank. Despite his mighty roars, there was no one in the vicinity to help. Exhausted and famished, the lion lay stranded without sustenance for several days.

Then, one day, a compassionate jackal happened upon the river. Observing the lion's dire predicament, the jackal approached cautiously and inquired, "What ails you? Why do you linger in this marsh?"

The lion recounted his woeful tale, and the jackal felt a surge of sympathy. Without hesitation, the jackal embarked on a mission to liberate the lion. With unwavering determination, he began to dislodge the clinging mud from the lion's trapped paw. But the lion, due to his prolonged immobility, found it challenging to free himself. Undeterred, the jackal laboured tirelessly, pulling, and pushing until the lion was at last extricated from his muddy prison. Grateful for his saviour's aid, the lion rejoiced in his newfound freedom, stretching his limbs with delight.

In gratitude for the jackal's selfless assistance, the lion extended an invitation for the jackal to dwell nearby, promising to share his bountiful catches whenever he hunted. The jackal accepted the lion's offer wholeheartedly, and together, they lived harmoniously, sharing their meals, and forging a profound friendship.

As time passed, both families expanded - the lion welcoming cubs, and the jackal, kid jackals. The offspring played gleefully together, oblivious to any differences. However, jealousy began to seep into the lioness's heart, for she perceived a threat in the close bond between her mate and the jackal.

She sternly admonished her cubs, urging them to shun the jackal's offspring, asserting, "They are not of our kind."

Turning to the lady jackal, the lioness declared, "Our families are growing, and it is no longer suitable for us to coexist. You should seek your own dwelling."

That night, the lady jackal conveyed the lioness's message to her mate, expressing her dismay. The jackal, upon hearing this, felt disheartened and confronted the lion, questioning, "If you no longer wish for us to remain, you should have expressed it directly to me. Why involve our families in this matter?"

The lion, genuinely remorseful, was ready to address the concerns with the lioness and find an impartial resolution. However, the shrewd jackal, acknowledging the lion's sincerity, expressed his gratitude for the lion's understanding. Instead of complicating the issue further, the jackal decided to find an alternative home for his family, deeming it the most prudent course of action for both clans.

While the lion felt a pang of sadness at their departure, he respected the jackal's choice and cherished the enduring bond of their friendship. Despite living separately, the lion and the jackal continued to meet daily, their friendship unwavering and enduring throughout their lifetimes.

The King's Finger

Long ago, there lived a king who enjoyed hunting. His faithful minister always accompanied him on his adventures.

One day, while hunting, the king shot a leopard. He went to check if the animal was dead. Suddenly, the leopard which was lying on the ground for dead sprang up and attacked him. The minister who was also a good hunter managed to kill the leopard.

He could not prevent his majesty the king from losing a finger. The king, visibly upset and angry, vented his frustration, "Why did I have to endure that leopard attack? Why did I have to lose my finger?"

The minister said calmly, "Good, bad, who knows?"

This annoyed the king more. The king who is known for his quick temper shouted, "You fool, you will pay for this. You are sentenced to life imprisonment."

"Your Majesty, good, bad, who knows?"

A few weeks later, when his wound was healed, the king was on the hunt again. He chased a deer, and no one managed to keep up with the king. He got separated from his attendants and got totally lost in the forest. He went into the deep forest where he was caught by the tribes who were preparing a sacrifice for their deity. Usually, they sacrifice animals. But today they were so excited to offer a king to please their goddess.

Suddenly the tribal chief shouted, "No! We can't offer an imperfect sacrifice to our deity. This man has only nine fingers! Sacrificing imperfect things will bring a curse on our tribe."

They let the king go. As he walked back to the palace he thought, "It is the loss of my finger that saved my life. He remembered the words of his minister, "Good, bad, who knows?"

As soon as he reached the palace, he ordered the release of the minister. The king exclaimed, "You are so wise, I wouldn't be alive today if I had all my fingers. It was a good thing that I lost my finger."

The minister replied calmly, "Your Majesty, good, bad, who knows?" He continued, "If you had not put me into jail, I would have been with you in that hunting expedition. I have all ten fingers. I would have been sacrificed in your place!"

Self-Reliance: A Herculean Twist

There was a hardworking farmer in a village. He had a big load of crops to take to the market. But heavy rains had turned the roads into muddy swamps. As he went down one of these muddy paths, his wagon's wheels got stuck in the mud. This made his horses unable to move the wagon. The farmer was very frustrated and got angry at his horses.

As he laboured to free the wagon, the wheels only sank deeper into the mud, and his horses grew wearier with every futile attempt. Disappointed and at a loss, the farmer remained seated in his cart, his weary horses looking at him with imploring eyes. Desperation mounted, and in his darkest moment, he reached out for divine assistance, invoking the god of strength, Hercules, to come to his aid. Much to his astonishment, Hercules appeared before him.

However, Hercules, though divine, was far from pleased with the scene before him - a farmer who, in his frustration, chose to rebuke his horses rather than act himself. Observing the weary beasts and the farmer's despair, Hercules implored, "Step down from your wagon and use your strength to push it. Do you believe that mere complaints and scolding your horses will free your wagon? I shall not offer my help until you've shown your determination to help yourself."

Duly humbled, the farmer obeyed Hercules's instruction. With every ounce of his might, he pushed against the wagon, his muscles straining. Slowly but surely, the wagon began to inch its way out of the muddy trap. The farmer, with unwavering determination, continued his efforts until, through his own hard work, he finally liberated the wagon from its muddy prison.

Grateful for the divine lesson and the guidance received, the farmer turned to express his gratitude to Hercules, only to find the god had quietly disappeared. Undaunted, the farmer coaxed his horses forward, and the wagon began its steady journey once more. As he continued his path to the market, he hummed a cheerful tune, carrying with him the invaluable lesson that, in life's trials, one must first exert one's own efforts before seeking help from others.

Curing the Mad King

Once upon a time, a kind-hearted king ruled over his kingdom with unwavering dedication to peace and justice. He tirelessly sought to do what was right and good for his subjects. Deep within the heart of the city lay a pristine well, the source of pure water from which both the king and his citizens quenched their thirst.

However, an envious enemy, burning with jealousy, hatched a malicious plan. Under the cover of night, this malevolent foe dispatched a powerful witch to poison the very well that had long sustained the people of the kingdom.

With the break of dawn, all the inhabitants of the city, except the king, partook of the well's water. The water had been tainted, and a dreadful madness seized those who drank from it. Chaos and confusion reigned as the once-rational citizens now clamoured, "Our king has gone insane!"

Concerned for their beloved ruler's well-being, the ministers and physicians convened meetings to discuss the treatment and potential recovery of the ailing king.

"I administered the most bitter herbs, but alas, they brought no solace to our king," mourned the herbalist.

"Have you contemplated the use of lizards' tongues and snakes' gallbladders?" inquired the wise minister.

"Thank you for reminding me of these rare remedies, and I shall certainly employ them in our king's treatment tomorrow," affirmed the physician.

The acupuncturist was next to share their efforts, stating, "I have inserted needles throughout the king's body, yet his condition remains unchanged."

"Perhaps," the minister recommended, "try needles of considerable thickness, resembling rods, for our dear king."

"I possess such needles in abundance, and I shall employ them on the king tomorrow," the diligent acupuncturist promised.

The following morning, the advisors and medical team stood prepared, armed with their ominous needles, formidable lizards' tongues, and treacherous snakes' gallbladders. When the king beheld this terrifying array of instruments, he could bear the prospect of such torment no longer. Voluntarily, he opted to drink from the tainted well, and soon he, too, descended into madness, aligning himself with his afflicted subjects.

The once-benevolent king soon became just as insane as everyone else, and the entire country returned to live in peace and harmony. The kingdom was once more united in their shared madness.

Questioning Traditions

Bella, the delightful 9-year-old girl, lived in a charming countryside home with her extended family. In this close-knit household lived her parents, grandparents, and her venerable great grandmother.

Bella possessed a heart full of politeness, an easy-going disposition, and a boundless love for those around her. Her insatiable curiosity led her to constantly seek out new knowledge, making her a vibrant and inquisitive soul.

One evening, Bella found herself in the kitchen, sitting beside her mother, who was engrossed in the art of cooking dinner. The centrepiece of the meal was a fish, and Bella watched in rapt fascination as her mother meticulously prepared it. With a deft hand, her mother removed the fish's head and tail before placing it into a baking pan.

Bella could not help but wonder why her mother had chosen to discard the fish's head and tail. Questions bubbled up within her, and she finally decided to voice them. Turning to her mother, she inquired, "Mom, why did you cut off the head and tail of the fish before baking?"

Her mother was taken aback by the unexpected query from her young daughter. The question had never crossed her mind before, and she paused to reflect. After a moment, she responded, "I've always done it that way."

Perplexed but undeterred, Bella pressed further, "But why, Mom?"

Her mother, with an affectionate smile, admitted, "Because this is how your grandmother has always done it, and I've followed her lead."

Bella's curiosity remained unsatisfied, but she decided to let the matter rest for the time being, planning to seek answers from her grandmother soon.

At the dinner table, as Bella observed her grandmother serving the same fish to everyone, she couldn't resist revisiting the topic. She questioned her grandmother about the practice of removing the head and tail before cooking.

Her grandmother paused, reflecting on the matter. She responded, "I don't know, dear. My mother always did it this way." With that, she continued serving the meal.

Bella found herself still searching for a deeper explanation. She understood that there had to be a reason, one beyond the responses her mother and grandmother had provided. However, she chose to eat her dinner in silence.

As bedtime approached, Bella kissed each family member goodnight and made her way to her great grandmother's room. Every night, Bella sought her great grandmother for bedtime stories and a goodnight kiss before retiring to her own bed.

Upon arriving in her great grandmother's room, Bella couldn't contain her curiosity and shared her burning question. To her delight, she was met with a wide smile from her great granny.

Unlike the others, her great grandmother had a swift and straightforward response. She replied with a chuckle, "My dear, it was because my baking pan was simply too small to accommodate the whole fish." Laughter filled the room as the generations shared a heartfelt moment of connection and understanding.

The Glass of Deception

In a quiet village near a forest, there lived a skilled hunter named David. He was well-known for his ability to track tricky animals through the dense woods. One sunny afternoon, he decided to invite his close friend, Michael, over for a relaxing evening of companionship and wine.

As the sun dipped below the horizon, the two friends found themselves seated in David's cosy cabin, sipping on glasses of fine red wine. The atmosphere was warm and inviting, the perfect setting for a peaceful evening. However, as they enjoyed their second glass of wine, something strange began to happen to Michael.

His demeanour changed drastically; a look of unease washed over him, and he suddenly grew pale. With an excuse about an early morning commitment, Michael abruptly left the gathering, leaving David bewildered by his friend's sudden departure.

Days passed, and David couldn't shake off the concern he felt for Michael. He decided to pay a visit to his friend to check on him. When he arrived at Michael's home, he found him looking tired and worried.

"Michael, my friend," David exclaimed with genuine concern, "What has happened to you? You look so tired and worried. Is there something troubling you?"

Michael hesitated for a moment, then finally admitted, "I don't know why, David, but ever since I drank your glass of wine that evening, I can hardly eat or sleep. I feel as if I'm on the brink of death."

David was taken aback by Michael's words, his thoughts racing to find an explanation for his friend's condition. Determined to uncover the truth, he decided to recreate the circumstances of that fateful evening. Returning to his cabin, he retrieved the same wine glass and poured a fresh glass of wine.

To his astonishment, as the wine settled in the glass, an image of a snake materialized inside it. The snake slithered and writhed, sending shivers down David's spine. He turned the glass this way and that, and the snake seemed to vanish, only to reappear when he shifted the glass again. David's face lit up with realization as he understood the source of his friend's distress.

It was not a real snake but merely a reflection of his bow hanging on the wall, playing tricks with the wine's refracted light. Relief washed over him as he cracked a wry smile at the absurdity of it all.

Filled with newfound understanding and armed with a plan, David quickly sent for Michael, inviting him back to the cabin. As Michael entered the cabin, he still looked anxious, fearing the worst.

David calmly poured wine into the same glass as before and watched as the "snake" once again appeared. Michael's eyes widened with fear at the sight, but David slowly turned the glass, revealing the bow's reflection. Michael's expression changed from fear to relief, a weight lifted from his shoulders as he realized he had not consumed a snake that fateful night.

Once the friend understood the source of his unfounded fear, he began to regain his appetite and sleep as before. The two friends shared a hearty laugh over the peculiar incident, cherishing their bond and the lesson learned.

Playing the Battle Drum

In a land ruled by a great and powerful king, there lived an extraordinary elephant named Gajendra. He was not an ordinary elephant; he was the king's most cherished companion and a symbol of strength and loyalty in the kingdom. Together, they had won numerous battles and expanded the kingdom's borders.

As the years passed, Gajendra grew older, but his spirit remained as strong as ever. The king, however, grew concerned for his beloved elephant's safety. He knew that Gajendra's age could make him vulnerable on the battlefield, and he could not bear the thought of losing his faithful friend.

One fateful day, the king made a difficult decision. He decreed that Gajendra would no longer be sent into battle. Instead, the wise and experienced elephant would remain in the royal compound, away from the dangers of war. Although the king believed he was protecting Gajendra, the elephant felt a deep sense of sorrow and longing for the thrilling adventures of the battlefield.

Weeks turned into months, and Gajendra's once-vibrant spirit began to wane. He grew listless and despondent, missing the company of his fellow soldiers and the exhilaration of victory. The king noticed Gajendra's melancholy and grew concerned for his friend's well-being.

One day, while Gajendra was near a tranquil lake, he ventured too close to the water's edge. The soft mud at the lake's edge proved treacherous, and the old elephant found himself sinking deeper and deeper into the mire. Panic seized Gajendra as he struggled in vain to free himself. His loud trumpeting echoed through the forest, reaching the ears of the nearby villagers.

News of Gajendra's predicament spread quickly, and soon a crowd gathered at the lake. The villagers and the king's soldiers tried every method they knew to rescue the trapped elephant. However, Gajendra's immense weight and the suction of the mud made it a nearly impossible task.

As despair settled over the crowd, an old, retired minister who had served the king for many years happened to pass by the lake. He approached the king and, after assessing the situation, offered an unconventional solution. The minister suggested playing the royal battle drums, a sound that had always filled the kingdom with courage and determination.

While the idea seemed strange and illogical to those present, the king trusted the wisdom of his old minister. With a sense of urgency, the drums were brought to the lake, and the skilled drummers began to play. The rhythmic, thunderous beats reverberated through the air, echoing in the hearts of all who listened.

To everyone's astonishment, including Gajendra himself, the elephant began to respond to the music. With each resounding beat, Gajendra found renewed strength and determination. He used his powerful legs to push against the mud, slowly pulling himself out of the treacherous swamp.

The king and the crowd watched in awe as Gajendra, driven by the spirit of the battle drums, emerged from the mire, victorious over the forces that had trapped him. The crowd erupted in cheers, and the king embraced his loyal friend.

When asked how he had known that the battle drums would work, the old minister smiled and explained that it was not a matter of physical ability but one of motivation and willpower. Gajendra had needed a reminder of his inner strength and the courage he had displayed on countless battlefields.

From that day forward, Gajendra's spirit never waned again. He continued to live in the royal compound, a symbol of strength and determination for the entire kingdom. The sound of the battle drums served as a reminder that true strength came from within, and with unwavering resolve, one could overcome even the most challenging obstacles.

The Unnoticed Guests

Once upon a time, in a thick forest, there was a tiny flea and his family who lived beneath a bush. They were always worried about their safety because the forest was full of big animals that could accidentally crush them. So, he decided to move with his family to a better and safer place.

One day, as the flea watched a massive elephant passing by, he got an idea. He saw how enormous and strong the elephant was and thought that if they could find refuge in the elephant's ear, they would be safe. So, he decided to propose this idea to the elephant.

The next day, as the elephant approached, the flea called out, "Hello, Mr. Elephant! My family and I want to move into your ear because it seems like the safest place for us. We'll give you a week to think about it and let us know if you have any concerns."

The elephant, not even aware of the tiny flea, continued walking, leaving the flea's proposal unanswered. Undeterred, the flea waited for a week, assuming the elephant's silence meant consent.

After a week, the flea and his family decided to make their move. As the elephant passed by again, they hopped onto his front leg and started climbing toward the elephant's ear. It took some time, but they finally found a comfortable spot on the big, floppy ear and settled in.

However, as time passed, they became uncomfortable in their new home. The little ones got sick, and Mrs. Flea wanted to find a healthier place to live. She asked her husband to talk to the elephant about it in a way that wouldn't hurt his feelings.

The flea approached the elephant carefully and said, "Sir, we've decided to move to a different neighbourhood. It's not about your ear, which is roomy and warm. It's just that my wife would like to live closer to her friends near the buffalo's foot. If this causes any inconvenience to you, please let us know within the week."

The elephant, who had never noticed the fleas in his ear, replied, "I didn't even know you were here, and I won't miss you when you leave."

So, one night, while the elephant was peacefully resting, the flea and his family quietly left the elephant's ear. They made their way down through the elephant's legs, leaving with a clear conscience, leaving behind a big, contented elephant who had no idea about the tiny visitors who had briefly stayed in his ear.

As the flea and his family left the elephant's ear, they shared a knowing glance, understanding that their brief sojourn had been a gentle reminder that sometimes, in the vast world around us, our importance is merely a fleeting thought in someone else's day.

The Bell on the Tiger's Neck

Once upon a time in ancient China, there was a legendary bell that hung around the neck of a fierce tiger. It was renowned for its strength and cunning, and its bell struck fear into the hearts of all who heard it. The villagers lived in constant dread of the tiger's presence, for it would roam the nearby forests, and its bell would ring ominously, signalling its proximity.

One day, a group of brave and wise monks from a nearby monastery gathered to discuss the predicament. They were renowned for their courage and wisdom, and they were determined to find a way to remove the bell from the tiger's neck, thus relieving the villagers of their fear. The monks contemplated various strategies.

As the discussion continued, a young monk named Wei stood up. He was known for his fearless spirit and unwavering determination. Wei said, "I am willing to face the tiger and attempt to remove the bell. But I will need your guidance and blessings to do so."

"What's your plan?" Master Liang, the chief monk asked.

"I would untie the bell while the tiger is in deep sleep," he replied.

"If I were given such a job," another monk, Feng announced, "I would feed the tiger first. If a tiger is full, he might not hurt people as much as he would when hungry. Then, I would try to approach him and untie the bell from his neck.

"None of your methods are absolutely safe," an old monk, Chang insisted. "The best way is to bind the tiger tightly with a strong rope. We must make sure the tiger can't hurt people. Then, I would go ahead and untie the bell."

“Who would bind the tiger?” asked a young monk, Chen. No one answered.

Finally, they went to the monk, their Spiritual Leader, Tai Chin, who was well known for solving difficult questions. After hearing the question and all the answers, Tai Chin laughed and declared, the person who put the bell on the tiger’s neck is the right person to take down the bell from the tiger’s neck. He’s the only one who knows how to do it well without risk.”

The Heart of a Mouse

Once upon a time, in a snug cottage on the countryside, there lived a small and timid mouse named Nakul. He was the pet of a kind magician who, although treated him well, could not remove Nakul's perpetual fear and sadness.

One cold winter night, Nakul, feeling scared and unsure, went looking for food. When he got to a dimly lit hallway, he saw Tommy, the scary housecat, slowly moving closer with a fierce look in his eyes. Nakul was really scared, but he managed to hide behind a big cupboard so Tommy couldn't see him.

Seeing Nakul's constant distress, the magician felt sorry for him and decided to turn him into a cat. As a cat named Raja, he enjoyed his new life, feeling happy, fearless, and making friends with other cats.

But one day, Raja's newfound happiness was shattered when Rustom, the neighbourhood dog, chased him. Fear once again overwhelmed him. The magician decided to change Raja, this time into a dog. Thus, Raja became Sultan, embracing his carefree life as a dog. Happy and fearless once again he was excited by the privileges of being a dog.

Sultan roamed the neighbourhood, even strolled away into unknown far-off places enjoying his newfound privileges, until he ventured into the deep forest. There, he encountered a real tiger, an encounter he had not anticipated. Running for his life, Sultan barely escaped, panting and trembling. He returned home, and the magician, out of sympathy, turned him into a mighty tiger, Bhima.

Now Bhima was a mighty and savage tiger. Fearless and ferocious, he started to live an aggressive life in the forest, hunting other animals. One day he witnessed a hunter killing another tiger who had chased him. Again, Bhima felt insecure and started worrying. He came complaining to the magician about the hunter. But this time, the magician got angry and refused to help him. Instead, the magician turned him back to the mouse who he once was.

With a sigh, the magician said to him, "Nothing I do for you is going to be of any help. Though you had the body of a tiger, you still have the heart of a mouse!"

So here he was, once again being Nakul, the timid, fearful little mouse of the magician's country cottage.

The Impostor

Once upon a time there lived many pigeons of different kinds and colours in the dovecote of a large farmyard. Out of them, grey-headed pigeons were many in number. The owner of the farm fed them regularly with a variety of grains. The pigeons were healthy, happy, and well-fed. They did not have to worry about food or shelter.

A jackdaw who lived near the farm with his flock used to notice the comfortable life of the pigeons. He was upset by the fact that the pigeons had such a relaxed and easy life. He enviously looked at the pigeons every day and wished to lead a happy life in the farmhouse. So, he made up a plan to steal the grains from the pigeons. He was determined about what he wanted and decided to disguise himself as one of the pigeons, to secure a share of the good things they enjoyed.

To camouflage into the flock of pigeons, he painted himself white from head to foot and quietly sneaked into the pigeon loft. The Jackdaw knew that if he kept quiet nobody would be able to identify him. And now the greedy jackdaw was able to enjoy the grains daily and live an easy-going life with the pigeons. This continued for several days. The pigeons never suspected him.

As the days passed the jackdaw became more and more comfortable living with the pigeons. One day, after a heavy feed, he thought, "These foolish pigeons will never know that I am not one of them." The thought made him burst into a loud laugh. He could not control himself. He started singing and instantly, the other pigeons noticed this strange voice.

They looked around and found the source of this weird voice and immediately realized that he was an impostor. In anger all the pigeons pecked the jackdaw so mercilessly that he began to bleed profusely.

The jackdaw flew away to save his life and went back straight to his own flock. But on seeing his painted body, they didn't recognize him. As he revealed his identity to his flock, the other jackdaws were furious and very disappointed about his dishonesty. The jackdaw realized his mistake and regretted his behaviour and apologized. They asked him to leave the flock and was forced to fly away from that place to save his life.

Now the jackdaw who wanted all the luxuries of life, became a homeless wanderer.

The Unfounded Dread

Vikram the lion was the king of the jungle. He was a very brave and mighty lion worthy of being the king. He was known for his great powers. Vikram had a strong-built body and legs, a large head with thick mane, a long tail and two glowing eyes which made him look ferocious. He was famous for his loud roar and hunting abilities. He lived with his pride in the deep dense forest and all the animals of the jungle were afraid of him.

Despite all these qualities, Vikram had kept one big secret which no one knew. He had a big problem. He was terrified by the crowing of the rooster. A chill would go down his spine whenever he heard a cock crowing.

He was very ashamed and unhappy about his fear and always thought, "Can life be worth living if the creature as mighty as me be afraid of the crowing of the rooster?"

He was tired of holding this secret for long. He used to cry bitterly when away from his pride. He tried numerous ways to overcome his fear but all in vain.

Vikram had a very dear and close friend in the jungle with whom he liked to spend time. It was Azad the friendly elephant. Azad was huge, and scary with a distinctly massive body, large ears, and long trunk. He was very intelligent with complex emotions, compassion, and self-awareness. Vikram was very fond of Azad as he was a happy-go-lucky elephant and seemed to have answers to all his concerns and apprehensions.

Finally, one day Vikram decided to share his long-kept secret with Azad, his best friend.

When Azad saw Vikran approaching with hesitation from a distance, he knew something was wrong.

"What troubles you, my friend? Can anything have the power to harm a beast so mighty and strong?" asked Azad as Vikram came closer.

Vikram with much reluctance confessed his secret in Azad's ears.

On hearing it, Azad started laughing loudly, "How can the king of the jungle be afraid of such a small creature? How can the crowing of a cock hurt you in any way?" "Just think about it!" and he continued laughing.

Vikram stood there confused because he did not expect this type of response from his best friend. Azad's laughter suddenly disappeared, and he started flapping his huge ears quickly and moved agitated and terrified.

Vikram could see that Azad was frightened out of his wits and asked, "What's wrong with you?"

"Oh really! You will be shocked to know. Do you see that black buzzing gnat? If he wishes to sting inside my ear, I'm doomed. I will go mad in pain."

Now it was Vikram's turn to be amused watching his enormous friend being petrified by a small gnat.

And as Azad managed to get rid of the gnat, the two friends laughed out loud.

After this incident, Vikram decided he would face his fear and see how quickly it disappears and not let any kind of fear come his way. He also understood that if we could see our fears through others' perspective, we would realize that most of our fears make no sense and that even the strongest creatures have some weak points. He built up his self-confidence once again. After all, he was the king of all the beasts!

Being of Great Value

Richmond farm was one of the biggest farms in the suburbs of Wilcox County. The farm belonged to Richard, who was a humble and hardworking farmer. He loved and took good care of all his animals. The farm was huge and was home to different kinds of livestock.

Milo, the little puppy, was the new addition to the farm. As Milo was very small and was the newest member of the farm, Richard always took him along wherever he went. Milo even accompanied him inside his home.

The other farm animals, observing their master's recently developed fondness for the young puppy, were not very appreciative.

One day little Milo decided to take a walk around his master's farm on his own. He slowly started his stroll around the stable where the horse was fed.

Suddenly he heard a strange loud noise calling out to him. As he looked around, he saw a big horse with oval-shaped hooves, long tails, long and slender legs, muscular and long thick neck, and large head looking down at him.

"You must be new here," the horse said.

Then he took a stride around little Milo inspecting him and said, "You see, I am the toughest and the most elegant animal around here."

The horse continued, "I suspect that an animal of your shape and size is of any value to the master at all. If you think you are the master's favourite, you will soon find out that the master loves me more than all the other animals because I carry large loads for him."

Little Milo looked at the horse, how enormous he was. “There is no point in arguing with such a huge animal,” he thought and quietly hung his head and walked away.

Just a few steps ahead he heard a mooing from an adjoining shed. When he looked up, he saw a four-legged white animal, with large black spots on its body. It had a big black nose and two sharp horns. It was a cow peeping out of its shed looking down at Milo.

“Holy crap!” said the cow who was all ears to the arrogant bragging of the horse.

“I have the most honoured position on this farm because the lady makes butter and cheese from my milk. And these are of great value in the market.”

“You, of course, provide nothing of value to the family.” The cow mocked Milo looking down at him scornfully.

“Cow, your position is no greater than mine,” came another voice from the nearby pen. Milo looked in the direction of the voice. It was the stocky-hooved sheep calling out.

“I lend the master wool to make his clothes. I provide warmth to the whole family. But you are correct about the dog. He gives nothing to the master. He is useless,” the sheep concluded.

One by one all the animals joined in the conversation, talking about their prominent position on the farm. The chicken asserted she produced eggs, and the cat how quick she was and how she kept the mice away from the house. All the animals did agree on one thing: that little Milo provided no service and so was of no value to the farm family.

Stung by the criticism of the other animals, Milo was depressed and disappeared into the dark and secluded place away from the animals and began to cry.

An old dog who was on his watch duty of the farm happened to pass by the dark corner where Milo was hiding. He heard Milo’s sobs and paused to listen to the little one’s story. “They are right,” Milo sobbed. “I provide no service to anyone.”

After listening to what Milo had to say, the old dog said, "It is true that you are too small to pull the wagon. And you will never produce eggs, milk, or wool. But it is foolish to cry about what you cannot do. You must use the ability the Creator gave you and try to bring value to people around you."

Milo was not convinced. He asked the older dog, "Of what value can I be to the Master? I am so small, and I have nothing to offer."

The old dog looked at little Milo with affection and said, "Little buddy! Go have a look at yourself and see how adorable you are. You can bring laughter and cheer around you. Isn't that a blessing? Think about it," and walked away into the darkness.

That night, when Richard came home exhausted from the fields, after long hours of hard work in the sun, little Milo ran to him. He licked his feet and jumped into his arms. Falling to the ground, the master and puppy romped on the floor.

Finally, holding him close to his chest and patting his head, Richard said, "No matter how tired I am when I get home, I feel better when you greet me and play with me. I wouldn't trade you for all the animals on the farm."

Heavenly Equality

Once upon a time, in a small village, there lived a poor farmer named John. John toiled day in and day out, cultivating his modest plot of land, hoping to make enough to feed his family and keep a roof over their heads. Life was a constant struggle, and luxury was an unfamiliar word in his vocabulary.

One fateful day, as the sun dipped below the horizon, John felt a sudden pain in his chest. He clutched his heart, gasping for breath, and then, in a moment, he was gone. His soul ascended to the gates of heaven, guided by a gentle breeze that carried him to his destination.

As John got closer to the Pearly Gates, he found himself standing next to a man who seemed incredibly rich and lavish. He was dressed in elegant clothes, adorned with sparkling jewellery that gleamed in the heavenly light. John, in his humble rags, felt a pang of insecurity. He couldn't resist comparing himself to the affluent stranger by his side.

In due time, Saint Peter, the guardian of the gates, appeared. He opened the gates and with a flourish, welcomed the rich man inside. John, unable to contain his curiosity, peeked through the gates to catch a glimpse of what awaited the man who had lived a life of affluence.

To his astonishment, a magnificent sight unfolded before John's eyes. A grand chorus of angels, dressed in resplendent white robes, greeted Saint Peter and the rich man with a rousing Bach chorale that seemed to resonate through the very heavens. The streets were alive with jubilant people who cheered and applauded. It was a scene of unimaginable splendour.

As the crowd's cheers subsided, Saint Peter addressed them, saying, "Welcome to the city of God. Make yourself at home."

With those words, the rich man walked down the golden streets, enveloped in the adoration of the heavenly throng.

When the rich man had disappeared, Saint Peter turned to John with a warm smile. "Welcome to the city of God," he said with equal enthusiasm. "Make yourself at home."

However, there were no angelic choruses singing in John's honour, and there were no crowds to greet him with applause. The stark contrast left John feeling disheartened and slighted. He could not help but voice his disappointment to Saint Peter.

"This is the last place I ever thought I would find discrimination," John said with a quiver in his voice. "All my life, I've watched the rich gain privileges that the poor were denied. I thought that when I came to the home of God, all would be equal. Yet when I enter the gates, I'm not greeted by either crowds or choruses."

Saint Peter, understanding John's concerns, gently placed a reassuring hand on his shoulder. "My dear friend," he began, "I can see how it appears that there is discrimination, but it's not true. Everything will be the same for you as for the rich man. You must understand that today is a special occasion. We receive poor farmers up here every day, but we haven't had a rich man in over 80 years."

With those words, John began to realize that the true treasures of heaven lay not in the grandeur of the welcome, but in the simple joys of eternal life in the presence of God. In the end, it was not the opulence that mattered, but the purity of one's heart and the love that transcended earthly riches.

The Three Fish

Once upon a time, deep within a dense forest, there was a small and tranquil lake. This serene oasis remained concealed from the nearby villagers, hidden amidst tall grass, thick bushes, and towering trees. In this peaceful haven, a variety of beautiful fish flourished, coexisting harmoniously.

Three fish, the closest of friends, called this lake their home. They had known each other since birth, although their personalities could not be more different.

The first fish was renowned for wisdom and sharpness. He carefully weighed the pros and cons before making decisions.

The second fish was intelligent, resourceful, and amiable. He relied on his quick thinking and clever solutions to navigate any challenges.

The third fish, however, was lazy and placed his faith in fate instead of hard work. His attitude often led him into trouble, and sometimes, his friends too.

One day, a group of fishermen from a nearby village ventured to the river in search of fish. No matter how hard they tried, they did not get even a single fish from the river as most of the fish had already moved to the lake. The fishermen toiled tirelessly throughout the day but found no success. Exhausted and disheartened, they were about to return home as night fell.

Suddenly, one of the fishermen spotted a group of cranes flying off with fish in their beaks. He cried out to his companions, "Look up there! Those birds! Where did they get those fish?"

Curious, the fishermen followed the flight path of the cranes into the dense forest. After wandering through the woods, they stumbled upon a lush area with tall trees and dense vegetation, and to their astonishment, they discovered the hidden lake teeming with fish. Their joy knew no bounds.

As night approached, the fishermen decided to return the next morning with more of their peers to seize the bountiful catch. They departed, brimming with excitement.

Observing the fishermen's arrival and overhearing their plans, the first fish grew anxious. He hurried to his friends, sharing the grim news.

"Oh no! This is a serious problem!" exclaimed the second fish. "What should we do?"

"I've made my decision. I'm leaving the lake through the canal, heading to a nearby river immediately. I'll return when things are safer. With all the village fishermen coming tomorrow, the river should be a more secure refuge," declared the first fish.

The second fish pondered his options. "I don't want to abandon my home. I'll devise a plan to save myself when the fishermen arrive."

Meanwhile, the third fish, true to his laid-back nature, shrugged off the concern. "I'm not leaving. I've lived here for so long, and I won't go anywhere. My luck has kept me safe until now, and I believe it'll continue to do so. Whatever will happen, will happen."

No matter how hard the first two fish tried to persuade their carefree friend, their efforts proved futile. In the end, they gave up.

That night, the first fish escaped through the canal into the nearby river.

The next morning, the village fishermen arrived and cast their nets into the lake, sending the fish into a frenzy. While the third fish remained passive, entrusting his fate to luck, the second fish desperately sought an escape plan.

Then, a stroke of luck: he spotted the lifeless body of an otter floating in the water. Seizing the opportunity, he swam inside the otter's carcass and stayed hidden for a while. When he emerged, he reeked of a decaying animal. Both the second and the third fish were ensnared in the net along with many others.

As the fishermen pulled in the net, the second fish played dead, keeping his eyes shut. A fisherman noticed him and, assuming he was lifeless, tossed him back into the lake, repulsed by the rotten smell. Once in the water, the second fish swiftly swam to safety.

In contrast, the third fish flailed helplessly in the net, unsure of what to do. His frantic movements only aggravated the fishermen, who struck him hard on the head, leaving him lifeless and destined for their catch.

A few days later, the first fish returned to the lake, and the second fish recounted the entire ordeal. "If only he had listened to us," the first fish lamented, "he'd be here with us today."

Together, they mourned the loss of their dear friend.

The Price of Soaring Friendship

Once upon a time, in a quiet pond, surrounded by lots of plants, lived a smart tortoise named Toris. This pond was a haven for various aquatic life, abundant with fish, aquatic plants, and vibrant weeds. It was also a gathering place for a variety of birds and animals who sought refuge, nourishment, and companionship. Among these frequent visitors were two geese who had formed a close friendship with Toris.

The two geese, whom Toris affectionately called Gilly and Gracie, visited the pond daily. They would gracefully glide through the clear water, catching fish, and basking in the tranquillity of the surroundings. Toris, known for his storytelling skill, would entertain them with tales of his adventures and share the wisdom he had gained over the years.

But, as fate would have it, a drought descended upon the land, casting a parched shadow over the once-green landscape. Months passed without rain, and the once-abundant pond began to shrink, its life swaying on the brink of extinction. The trees withered, the greenery faded, and the fish vanished. Birds and animals, unable to find sustenance, perished or migrated in search of a more hospitable place to survive.

The geese, concerned for their own well-being, decided that they too must leave the desolate pond in search of greener pastures, quite literally. They knew that Toris, who could not fly, faced an agonizing fate if left behind.

One morning, Gilly and Gracie approached Toris with heavy hearts, preparing to break the news of their departure. Toris, with a hint of dread, inquired, "What? Where are you going?"

Gracie gently replied, "Oh, Toris, there is no food left here, and the pond has withered away. We are flying to a place where ponds are brimming with water and teeming with fish, surrounded by lush, green plants."

Toris, his voice quivering with emotion, implored, "But what will happen to me? Are you going to leave me here to wither away and die? There are no plants or leaves in this dried-up pond to sustain me. Please, take me with you. Am I not your friend?"

The geese were in a dilemma. They couldn't bear the thought of abandoning their dear friend to such a cruel fate. They pondered deeply on how they could help Toris, who was incapable of flight, to begin this journey.

"How can we help you?" Gilly inquired, with great concern. "You can't fly like us. We can fly great distances in the sky, but you can't. How will you come with us?"

The trio sat by the dry pond, racking their brains for a solution. Then, after a thoughtful pause, his eyes lit up with a glimmer of hope.

"I've got it! I know how I can come with you," Toris exclaimed.

Both geese leaned in, their curiosity awakened. "How, Toris? Please, tell us," they urged.

"Fetch me a stick, a long and sturdy one, but thin enough for me to grip firmly with my mouth," Toris instructed.

Without hesitation, the geese took to the air and soon returned with the requested stick.

"Watch closely," Toris explained, "I will bite the stick in the middle and hold it securely in my mouth. Each of you will grasp one end of the stick, and we will fly together. Our problem is solved."

With hope in their hearts and determination in their eyes, Gilly and Gracie took their positions, each gripping an end of the stick securely in their beaks. Toris clamped down on the middle, and with a unified effort, they took to the sky.

As they ascended higher and higher, they passed over sprawling fields, rolling hills, and even a bustling city. Onlookers gazed in astonishment at the extraordinary sight of two geese carrying a tortoise on a stick. People marvelled at this wondrous display, clapping their hands, and shouting in amazement.

Children, especially, were enthralled by the sight. They cheered gleefully, "Look! A tortoise in the sky! Look up there!"

Toris, not quite appreciating the attention, decided to respond to the enthusiastic crowd. With great effort, he opened his mouth to address the onlookers. However, the moment he did, his grip on the stick weakened, and he tumbled to the ground below.

Gilly and Gracie quickly descended to check on their fallen friend, but their hearts sank when they realized that Toris lay motionless on a rock. He had sacrificed himself to communicate with the people, and it had cost him his life.

With heavy hearts, the geese, now burdened with grief, realized that they could do nothing more for their beloved friend. They took to the skies once again, this time with an empty place in their hearts, leaving behind a fallen hero who had dared to defy the odds to be with his friends till the very end.

The Sweetness of the Strawberry

In a dense forest filled with ancient trees and untamed secrets, a man walked alone, seeking solace in the beauty of nature.

As he ventured deeper into the woods, he stumbled upon a scene that sent a shiver down his spine - a majestic tiger, poised to pounce, its amber eyes locked onto him. Fear surged through his veins. In a desperate bid for survival, he summoned every ounce of strength and began to sprint, the tiger's predatory instincts hot on his heels.

The relentless chase continued, with the man's heart pounding in his chest and his breath ragged. He sprinted further and further into the wilderness, but the relentless pursuit of the tiger persisted. The man's heart sank as he realized that he was running out of options.

Finally, he reached the edge of a cliff, the abyss of certain doom yawning below. With no other recourse, he clung to a sturdy vine, his fingers gripping it with a desperate tenacity. Just as the ferocious tiger pounced at him, he managed to haul himself over the precipice's edge, escaping the jaws of death by the narrowest of margins.

Yet, his respite was short-lived. As he scrambled for safety on the precipice, he lost his footing and began to plunge into the abyss below. In his free fall, his hand snatched at the only lifeline available - a small strawberry bush clinging to the side of the precipice.

But the man's relief was short-lived as he realized that the roots of the fragile bush were giving way. He hung precariously, swaying between the gaping mouth of the chasm below, which seemed ready to devour him, and the ferocious tiger above, eager to claim his life.

His heart raced, and terror gripped him as he dangled there, suspended between two imminent deaths. He could feel the warmth of the tiger's breath above and the cold abyss beckoning below. It was a chilling, inescapable predicament.

Amid his harrowing ordeal, his eyes fell upon a sight that defied the looming terror - a single, ripe strawberry, glistening with the promise of sweetness, nestled amid the leaves of the bush he clung to. It hung there, a fragile symbol of life's simple pleasures amidst overwhelming peril.

Grasping the bush with one hand, he plucked the strawberry with the other and put it into his mouth. Never in his life had a strawberry tasted so sweet! He experienced a moment of profound clarity. In that suspended moment, he realized that the sweetness of life could be savoured even in the direst of circumstances.

Shared Kindness

In a quiet village there lived two brothers who were the fortunate heirs to their father's bountiful estate. Together, they diligently worked the vast land, sharing the toils and the rewards of their labour, their brotherly bond unbreakable.

One night, the elder brother awoke from a deep slumber, his thoughts heavy with contemplation. He pondered the circumstances of his life and his brother's. "My dear brother," he mused, "is now married and has children to care for. He carries the weight of responsibilities and expenses that do not burden me. I, on the other hand, have only myself to provide for. It is only fair that I ease his anxieties."

With a heart full of goodwill, he silently rose from his bed and collected sacks filled with the golden corn they had sown together. These sacks he carried to his brother's granary, driven by the spirit of generosity and fairness. Once his task was complete, he returned to his own bed, content with his actions.

Shortly thereafter, the younger brother also awoke in the depths of the night. He too contemplated their shared inheritance and the divergent paths their lives had taken. "My beloved brother," he reflected, "has chosen to remain unmarried, free from the responsibilities of a family. I, on the other hand, have the joys of a loving spouse and several children. While my life is filled with security, who will care for my dear brother as he grows old?"

With a selfless heart, he rose from his slumber and gathered sacks filled with the golden corn. Silently, he transported them to his brother's granary, guided by the principles of fairness and brotherly love. His heart was light, knowing that he could contribute in some small way to his brother's well-being.

The following morning, both brothers awoke to a surprising sight. They were astonished to find that the quantity of corn in their respective granaries remained unchanged. Perplexed, they could not comprehend why, despite their nocturnal efforts, the quantities seemed to remain constant.

Year after year, the same mysterious occurrence transpired, leaving them both bewildered. Each brother was left to wonder how the corn continued to balance itself between their granaries, seemingly untouched by their acts of kindness.

One fateful night, a moment of realization descended upon them simultaneously. A deep resolve overcame each of them, and they were more determined than ever to carry out their act of fraternal generosity. They gathered larger piles of corn, dragging them through the moonlit fields and up the hill to each other's barns.

At the pinnacle of the hill, bathed in the gentle glow of the moon, they unexpectedly crossed paths. The truth struck them like a bolt of lightning. They dropped their sacks of corn and, without uttering a single word, embraced each other tightly. In that embrace, they discovered the extraordinary depth of their brotherly love and the profound symmetry of their actions.

The Country Feast, the City Fears

Once upon a time, a city mouse decided to visit his cousin in the countryside. After a long journey, he arrived in a village. This was his first experience of the countryside. He was excited to see the calm surroundings, the fields, the trees, and farms. But as he started strolling around the village he passed across some cattle farms, barns, and wet compost areas. He wondered, "This place's so dirty and smells awful. I wonder how someone lives here! Life in the city is much better."

After a long and tiring scampering, the city mouse finally met his cousin. They spent a lot of time together. The country mouse showed his new friend all around the countryside and let him taste all the nuts and fruits of the land.

It was almost night-time, so the country mouse invited the town mouse to his humble home for dinner. Together they headed towards the country mouse's cottage. The country mouse had made himself a house underground. The city mouse was shocked to see the inside of this home. The place was a dark and airless little hole. He thought, "How does my cousin live in such a noxious place?"

On the other hand, the country mouse was very excited to have his cousin from the city for dinner at his place. They sat down to eat a dinner of fresh peas, acorns, corn, fresh carrots, sweet potatoes, and plenty of fresh water to drink. The city mouse ate very sparingly as the food wasn't nearly as exciting as his town meal. He nibbled a little of this and a little of that, and by his manner one could make out that the city mouse found the food very bland and tasteless. It was evident that he ate that simple food only to be polite.

Meanwhile, during dinner, the city mouse kept on talking about his life in the city. "The city is amazing. I go to the theatre, the museum, the mall and to the restaurants. My house has so many rooms. You absolutely must visit my place. Have you ever tried Chinese cuisine, or the Italian? In the city you get to taste all kinds of food from around the world." And so, he went on and on describing the luxuries of the town.

After spending long hours talking, they slept quietly on the comfy beds in the burrow. All night the country mouse thought about the luxurious life of his cousin in the town. And the next morning when the city mouse asked the country mouse to accompany him to the town, he was more than excited. The country mouse quickly packed his bag, and the two friends started their journey towards the town.

The country mouse was thrilled to embark on his adventure. As they approached the city he was delighted and at the same time amazed by what he saw around. Tall buildings, shiny cars, and their loud noise, colourful people, busy restaurants, delicious smelling food. Everything was so different here and at the same time a bit strange too.

After a long journey, they finally arrived at the city mouse's house. It was indeed an enormous house, beautifully decorated with exquisite furniture. On the table they found a variety of food: jellies, pastries, cheese, fruits, and bread: indeed, the most tempting food that a mouse could imagine. The country mouse was really impressed by the lifestyle of his cousin and decided to stay with him in the city.

But just as the country mouse made up his mind and was about to nibble a dainty bit of pastry, a cat with sharp teeth and sharp claws jumped up on the table. The city cat chased the two mice off the table and across the floor. In great fear the mice scurried to a hiding place. They escaped just in time dashing through a small hole in the wall, where they lay quite still for a long time, hardly daring to breathe until the cat had gone. When at last they ventured back to the feast, to their disappointment all the food was gone. And, then the door opened and in came the servants to clear the table followed by their dog. The mice ran once again breathless to the little hole, before the dog could spot them.

In the hiding place city mouse whispered in his cousin's ear, "That cat killed my father and my mother. Now I'm frequently the target of his attack." On hearing this, the country mouse decided to go back.

The country mouse collected his things and scampered out of the big house. As he was leaving, he shouted out, "There are too many dangers here! I'd rather eat my humble food in peace and silence than eat your delicious food surrounded by danger!

He added, "It is better to gnaw the bean I get in the country than be gnawed by fear in the city!"

The Animal School

Once upon a time in a dense and vast forest, there stood a remarkable institution known as "The Animal School." It was no ordinary school, for it was exclusively attended by the young animals of the forest. Here, they learned valuable lessons about life, survival, and the importance of love, and unity in diversity.

The principal of the school was a wise old elephant, Guru Ganesh. With his deep knowledge of the forest and its inhabitants, Ganesh was respected by all the young animals. Under the shade of a giant oak tree, he taught his students not only the essential skills of survival but also the values of cooperation, respect, and empathy.

One sunny morning, Guru Ganesh gathered his students beneath an ancient banyan tree and spoke of the importance of cooperation in the forest. "Dear students," he began, "today, I have a special challenge for you. I want each of you to work together with a partner from a different species to create something wonderful for our forest. Each of you will be paired with an animal from a different species, and together, you will embark on a project that will benefit our entire forest community."

The young animals were filled with enthusiasm and curiosity. They paired up, and soon the forest was buzzing with activity. Squirrel teamed up with rabbit, deer with fox, and bird with hedgehog.

Abu the monkey partnered with Arjun the peacock to create a vibrant garden filled with colourful flowers and fruit-bearing trees, attracting a wide variety of birds, insects, and other creatures.

Punit the mongoose and Seshu the snake joined forces to clear a safe path through the dense underbrush, allowing animals to move through the forest more easily without fear of encounters with dangerous predators.

Hari the lion and Arun the deer decided to organize a grand cultural festival, where animals from all species showcased their unique talents in music, dance, and storytelling. The festival became an annual tradition that brought the entire forest community together in celebration.

In a quiet corner of the forest, Vivek the tortoise and Amir the rabbit collaborated on constructing a peaceful meditation garden, where animals could come to find solace, reflect, and find inner peace.

As the projects neared completion, the animals gathered to share their accomplishments. The forest had transformed into a harmonious community with a thriving garden, a safe path, a vibrant cultural festival, and a serene meditation space. The animals knew that they had created something special that would benefit them all.

Guru Ganesh, observing his students and the thriving forest community they had built, could not have been prouder. He knew that his role was not only to teach survival skills but also to instil the values of collaboration, empathy, and care for one another, reflecting the ancient wisdom and diversity of India.

In the heart of the forest, the animals celebrated their achievements with a grand festival filled with music, dance, and delicious food from different corners of the jungle. And so, The Animal School taught its students that the most important lessons in life were not just about individual survival but about the strength and beauty that could emerge when animals of different species came together in harmony, unity, and care for their shared home.

Silent Acts of Heroism

Once there lived a rich lord who held sway over vast domains, who was the proud owner of a majestic falcon. This falcon was no ordinary bird; it was a creature of grace and obedience, trained meticulously to heed its master's every command.

One fine day, the lord went on a hunting expedition, accompanied by a retinue of loyal servants. As they ventured deep into the wild, the falcon's keen eyes spotted its quarry, and it swiftly caught its first rabbit of the day. The lord, filled with admiration and pride for his magnificent companion, affectionately stroked the bird's feathers.

However, as the day wore on, the lord felt very thirsty. His quest for water led him to a serene hillside, where a meagre spring came into view, its water trickling slowly from the earth, drop by precious drop. With a sense of relief, the lord cupped his hands under the spring, waiting patiently for the cup to fill.

Just as he was about to quench his thirst, the falcon flapped its powerful wings and, in an act that seemed both strange and untimely, spilled the water. The lord, taken aback, scolded the bird for its seemingly reckless action. Undeterred, he again placed the cup beneath the spring, this time exercising even greater patience. But once more, the falcon landed upon his wrist, causing the water to spill to the ground. The lord's scolding grew harsher.

Determined to satisfy his thirst, the lord filled his cup a third time. Yet again, the falcon intervened, interrupting his drink. This time, the lord's anger flared into a tempestuous rage, and he struck the bird with a forceful blow that ended its life in an instant.

With the falcon lying lifeless before him, the lord's anger gave way to profound sadness. It was then that he instructed a servant to draw water from the spring. The servant returned, however, with an empty cup and a grave expression.

"Sir," the servant began, "I regret to inform you that the water from the spring is not fit to drink. There is a snake dwelling within its depths, and it has poisoned the water. Had you drunk from the spring, you would have met a grave fate."

The lord's head hung low, and his eyes welled with tears. "The falcon," he lamented, "in its wisdom, saved my life. Yet, in a moment of anger, I have brought about its untimely end."

In that poignant moment of realization, the lord understood the profound sacrifice the falcon had made, sparing his life without the ability to convey the imminent danger. With remorse weighing heavily upon his heart, he mourned the loss of his loyal companion, whose final act had been one of unspoken heroism and selflessness.

Grandpa's Final Lessons

High on a hill overlooking the tranquil bay, Grandpa and his curious grandson, Johnny, stood together on the balcony of their family home. Down below, they watched the diligent fishermen as they tended to their boats, cleaning and preparing for their next voyage. Johnny, with a mind full of questions, knew that his Grandpa held the answers to life's mysteries, especially this time when the questions were about the man himself.

Johnny's mother had solemnly whispered to him, "Grandpa's coming home to die."

"But everyone said…" Johnny began, his voice trembling with uncertainty.

"I know," Grandpa replied, his eyes filled with understanding.

Later, when the two were alone, away from prying ears, Grandpa gently broached the subject. "Johnny, your mom told you about me, didn't she?"

Johnny nodded, his young face showing great concern.

"It's nothing to be afraid of, my boy," Grandpa assured him, his voice tender and reassuring. "It's a part of life, and it happens to all of us eventually."

Johnny, his curiosity burning, asked the question that weighed heavily on his heart. "Are you scared, Grandpa?"

Grandpa paused, his eyes reflecting a mixture of emotions. "Yes, Johnny, I am. It's a bit like the feeling you get when you're on those thrilling rides at the amusement park. You know there's an element of uncertainty, but deep down, you believe that everything will turn out all right."

The young boy's eyes welled with concern. "Does it hurt, Grandpa?"

A small smile touched Grandpa's lips as he responded, "Being dead doesn't hurt, Johnny. It's the journey to that point that can be painful sometimes. But don't worry, I've got my pills to help me along."

In the days that followed, as Grandpa's strength waned, he and Johnny shared precious moments together. Grandpa learned to play video games, taught Johnny multiplication tables, and showed him how to bear the pain while balancing the various medications.

"It's quite unfortunate," lamented Grandpa's oldest friend, shaking his head. "But what can we do?"

Grandpa, despite his weakening condition, maintained a cheerful disposition, especially when he was with Johnny. They remained pals, even as their time together grew shorter.

As the end drew near, Johnny could not help but ask the question that weighed on his heart. "But why do you have to die, Grandpa?"

Grandpa, gazing at his beloved grandson, offered an explanation filled with tenderness and wisdom. "You know how it is when you get so sleepy that you can't keep your eyes open, and no matter how much fun you're having, you still have to sleep?"

Johnny nodded, his eyes reflecting his understanding.

"Well," Grandpa continued, "that's kind of what I think dying is like, Johnny. I just can't keep my eyes open any more. Just like you need your sleep, I need mine."

In that moment of shared understanding and love, Grandpa's wisdom comforted Johnny, and the fear of the unknown began to fade away.

Winning the Robbers' Hearts

Long ago, a young boy embarked on a journey to Baghdad, crossing the vast desert. His mother had sewn a handful of gold coins into his blanket, instructing him to keep it safe and not to reveal its contents until he reached the city. This precautionary measure was taken to protect him from potential robbers, as in those days, there were no trains, cars, or caravans. Travelers had to rely on their own two feet and wits to navigate the perilous desert.

As the young lad moved further into the desert, he encountered a group of robbers. Although they assumed that a little boy like him would not possess a substantial sum of money, they decided to ask anyway, "Do you have any coins, gold, or silver?"

Inherently honest, the boy felt compelled to speak the truth and replied, "Yes, I do." His conscience would not allow him to do otherwise.

Curious, the robbers inquired further, "Where are these coins?"

With firm honesty, the boy responded, "They are sewn into this very blanket."

Remarkably, the act of the boy disclosing the location of his hidden wealth had an unexpected effect on the robbers. Instead of succumbing to their criminal inclinations, their hearts were touched by the boy's honesty. They realized that the young traveller posed no threat and had voluntarily revealed his secret.

In a surprising turn of events, the robbers decided to act justly. They declared, "We might have resorted to theft had you not been truthful with us." With newfound integrity, they chose to release the boy unharmed and without pilfering his precious coins.

The Quest for the Buddha

Once upon a time, a young man, fuelled by a fervent desire for knowledge, left behind the familiar surroundings of school and the warmth of his parents' embrace. His quest was to traverse the world and seek enlightenment from a learned teacher residing beyond the formidable mountains. With eager anticipation, he packed his belongings and embarked on a journey fraught with the promise of wisdom.

As he trod the path of self-discovery, fate intervened in the form of a storm that ambushed him in the foothills. Seeking refuge, the young man spotted a wisp of smoke emanating from a fire and drawn to its warmth, found himself before a small hut. There, in a tree, sat a venerable hermit - an embodiment of holiness.

Curiosity burning within him, the hermit posed a question to the young seeker, "What are you seeking?"

With unwavering determination, the youth replied, “Understanding the Buddha mind.” The hermit, a wellspring of sagacity, offered an alternative perspective, “Why squander your time with a renowned teacher when I can connect you with a genuine Buddha—a truly enlightened person?”

Intrigued, the young man inquired, “Would you?”

The hermit, with serene confidence, imparted his guidance, “Go out in the morning, retrace your steps along the path you have come. When you see a person wearing a blanket over the shoulders, carrying a lamp with shoes on the wrong feet, you will know this is your Buddha, and you ought to take your wisdom from this enlightened person”

Emboldened by the prospect of encountering such a being, the young man eagerly set forth at dawn. His quest led him to scrutinize every person he encountered, studying their feet and attire. Despite two days and one night of relentless searching, he found himself under the moonlit sky, back on his own doorstep."

Filled with urgency, he beat on the door and cried, "Let me in! It's me."

To his bewilderment, a voice from within responded, "Oh, I'm glad you're home, honey."

In a moment of profound revelation, the young man discovered his mother greeting him. Carrying a lamp and wearing a blanket thrown over her nightgown; in her haste, she had mistakenly donned the wrong slippers on the wrong feet. The irony of enlightenment unfolding at his own doorstep left the young seeker in awe, realizing that wisdom could be found in the ordinary, even within the loving embrace of his home.

❑❑

BIBLIOGRAPHY

Boyd, Brian. (2010). *On the Origin of Stories; Evolution, Cognition and Fiction.* Cambridge: Harvard Univ.

Burns, G. W. (2005). *101 Healing Stories for Children and Teens: Using Metaphors in Therapy*. New York: Wiley.

Coleridge, ST. (1984). *Biographia Literaria*. (ed. James Engell and Jackson Bate). New Jersey: Princeton Univ.

Gersie, Alida & King, Nancy. (1989). *Storymaking in Education and Therapy*. London: Jessica Kingsley.

Gottschall, Jonathan. (2013). *The Storytelling Animal: How Stories Make Us Human*. Boston: Mariner.

Karnad, Girish (1988). *Nagamandala.* New Delhi: OUP.

Perrow, Susan. (2018). *Healing Stories for Challenging Behaviour.* New York: Hawthorn.

Stone, Richard. (2004). *The Healing Art of Storytelling*. New York: Authors Choice.

White, W.R. (1986). *Stories For Telling: A Treasury for Christian Storytellers*. Minneapolis: Augsburg.

Wright, Andrew (2009), Storytelling with Children (2nd ed.), Oxford: Oxford University Press.

http://assets.press.princeton.edu/chapters/s9676.pdf

https://news.artnet.com/art-world/chauvet-cave-paintings-404753
https://www.psychologytoday.com/us/blog/owning-pink/201211/the-healing-power-telling-your-story
https://medium.com/swlh/how-storytelling-helps-you-heal-9ed3290c95a9